HANDBOOK OF VACUUM PHYSICS

VOLUME 2

PHYSICAL ELECTRONICS

PARTS 4, 5 and 6

HANDBOOK OF VACUUM PHYSICS

VOLUME 2

PHYSICAL ELECTRONICS

EDITED BY

A. H. BECK

Engineering Laboratory, Cambridge

Part 4

E. A. Ash—Electron Optics

Part 5

A. H. Beck—Heavy Current Electron Beams

Part 6

A. H. Beck—Gridded Valves. Theory and Design Principles

PERGAMON PRESS

OXFORD · LONDON · EDINBURGH · NEW YORK

TORONTO · SYDNEY · PARIS · BRAUNSCHWEIG

Pergamon Press Ltd., Headington Hill Hall, Oxford
4 & 5 Fitzroy Square, London W.1
Pergamon Press (Scotland) Ltd., 2 & 3 Teviot Place, Edinburgh 1
Pergamon Press Inc., 44–01 21st Street, Long Island City, New York 11101
Pergamon of Canada Ltd., 207 Queen's Quay West, Toronto 1
Pergamon Press (Aust.) Pty. Ltd., 19a Boundary Street, Rushcutters Bay, N.S.W. 2011, Australia
Pergamon Press S.A.R.L., 24 rue des Écoles, Paris 5e
Vieweg & Sohn GmbH, Burgplatz 1, Braunschweig

First edition 1968

Library of Congress Catalog Card No. 63–21443

PRINTED IN GREAT BRITAIN BY
ALDEN & MOWBRAY LTD AT THE ALDEN PRESS, OXFORD
08 012440 2

CONTENTS

PART 4

ELECTRON OPTICS

E. A. Ash

Department of Electrical Engineering,
University College, London

PART 4

ELECTRON OPTICS

E. A. Ash

Department of Electrical Engineering,
University College, London

1. INTRODUCTION

The operation of an important class of electronic devices can be described in terms which are very similar to those used in classical light optics. The analogy is deep, in that both sciences flow from a common fundamental principle, and at the same time is conceptually extremely useful. There are naturally limits to how far the parallel can be drawn; one of these limits is set by the need to take the mutual repulsion of the electrons into account, when the current density is sufficiently large. Although, even in this event, the picture could, if desired, be stretched—one would have to talk about a current dependent refractive index—it is of very dubious value to do so. At the same time it so happens that most devices in which this effect would be important are distinguished also from the low current devices by the nature of the applications. The discussion of high current beams is presented in a companion article.

Another limit to the parallel with classical optics is encountered if one wishes to treat time-varying fields by a generalization of electron optics[1]. Although mathematically the two subjects are contiguous, the devices to which they relate are sharply differentiated in function; the present article will be confined to time-independent situations.

Although the science of electron optics has spread in so many directions that it is not now easy to delineate its frontiers, its origin is clearly defined, and is to be found in a single paper—that by Busch[2,3], which first explained the action of "concentrating" coils in cathode ray oscilloscopes. Even before this time there was a great deal of existing art and theory of electron dynamics. However, the concept of an electron *lens* is, it is generally agreed, to be traced to this paper. The concept at once stimulated further experiments which in only six years led to an electron microscope which surpassed the performance of the best optical microscope then available. The fascinating early history of the electron microscope has been recorded by Gabor[(4)] and the reader with a taste for the history of the subject is referred to this article.

Electron optics shares with light optics the characteristics that it is amenable to analytical treatment, without the need to make drastic assumptions limiting the validity of the results. Most of the important properties of electron optical systems can be analysed starting with only three equations—Newton's and Laplace's equations and the Lorentz force law—and the applied mathematician need not burden himself with any further knowledge of physics. This circumstance has attracted a great deal of analytical effort; in a brief survey it is therefore very pertinent to ask to what extent has this theoretical attack contributed to the performance of electron optical instruments. The answer, to those who have not succeeded in eliminating a utilitarian streak from their natures, must be depressing. Indeed, in considering the electron microscope one might speculate as to whether any other invention has produced such a high ratio of theoretical expressions propounded to expressions used in grim earnest. This is not to deny the elegance and beauty of the mature theory of electron optics, nor the use which might be made of it in the future, possibly in an entirely different context. However, for the purpose of this survey, as far as possible we will adopt the rule of giving only such analysis as is very directly related to the design or understanding of electron optical devices.

At this point we must raise another issue, and one which will encourage us to adopt a slightly different point of view from that taken by previous and longer treatments. How does the existence of the high speed computer affect the kind of analysis which one needs in order to obtain results by the quickest route? It is one of the stranger doctrines of applied mathematics that once a differential equation has been transformed to a definite integral the problem has been "solved".

The "solution" is regarded as only a little tarnished if the integral contains parameters which represent the solution of auxiliary equations. On the other hand, the user has to decide whether it is quicker to compute the "solution", or to effect a direct integration of the starting equations. To be useful, an analytical solution must satisfy one of three possible requirements:

(i) It should be of use in proving general theorems.
(ii) It should require fewer arithmetical operations in order to attain a numerical result.
(iii) It should be so simple as to be meaningful and suggestive by inspection.

It is our impression that this set of criteria means that in many cases we are better advised to hand the raw material to the computer rather than to attempt any prior cooking.†

In a short survey it is clear that we can only include the most basic results. However, it is hoped that enough can be included to enable the reader to gain a general understanding of the principles underlying electron optical devices and their design. The equations have been written in the non-relativistic form. This approximation is adequate for most of the applications discussed, but is not usable for some special electron microscopes such as those which use voltages up to 500 kV[(5)]. In most cases the modifications required are simple to effect. For a detailed treatment the reader is referred to L. de Broglie's book[(6)].

In most of the discussion the electron will be regarded as a particle, the wave description being referred to only where we are directly concerned with wave phenomena, notably in discussing diffraction effects in the electron microscope. The reason is simply that most of the results required are adequately reflected in the corpuscular theory.

It will be assumed that the reader is familiar with the electrostatic and magnetostatic theory and that in some context he has encountered the Lorentz force F, which in the M.K.S. units, used throughout, is

$$F = e(\boldsymbol{E}+\boldsymbol{v}\times\boldsymbol{B}). \tag{1.1}$$

† The advent of the computer has, of course, no effect on which of two formulations leads more rapidly to a numerical result. The fact that, in the pre-computer era, it had in many cases not been noticed that the analytical result was no nearer the solution than the starting point suggests that neither had ever been computed.

2. PARAXIAL THEORY OF ELECTROSTATIC AND MAGNETIC LENSES

2.1. General considerations

The principal concern of the theory of electron optics is the demonstration and evaluation of the imaging properties of rotationally symmetric electric and magnetic fields. We will see that the action of such fields on electron streams is very closely analogous to that of lens systems on beams of light. The analogy with light imaging is sufficiently close to allow us to use much of the same terminology, not only with regard to the Gaussian imaging but also with regard to the aberrations.

The phenomenon of image formation by particles travelling in a system having circular symmetry is quite general; it is by no means

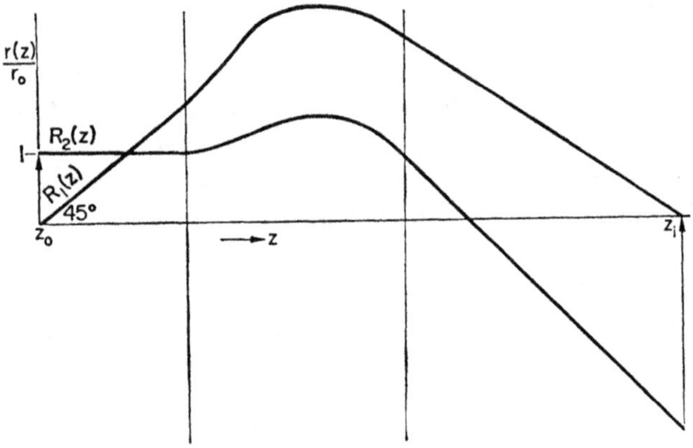

FIG. 1. Two particular solutions of eq. (2.4).

confined to the case of charged particles in electrostatic or magnetostatic fields. Neutral particles travelling in a rotationally symmetric gravitational field, or compressible fluids, represent other examples.

With reference to Fig. 1 consider a particle starting from the symmetry axis at $z = z_0$ moving through the force field, and intersecting the symmetry axis for a second time at $z = z_i$.† Let the radial

† This implies a converging lens; the proof is readily extended to diverging lenses. However, as will appear in section 2.5, divergent electron lenses in which object and image are outside the force field do not exist.

force experienced by the particle be $F_r(r, z)$. Suppose we now expand F_r in powers of r:

$$F_r(r, z) = \sum_{n=0}^{\infty} G_n(z)\, r^n. \tag{2.1}$$

The restraint imposed by circular symmetry immediately requires that all even powers of r must vanish, and the first term in the expansion is $G_1(z)r$. For sufficiently small values of r the higher terms can be neglected, and we may write the radial equation of motion in the form,

$$m\ddot{r} = G_1(z)\, r, \tag{2.2}$$

where m is the mass of the particle. Let the particle velocity on the axis be $v(z)$. At a small distance r from the axis the velocity will differ from $v(z)$ on two counts: firstly, due to a difference in potential energy arising from the radial force field, secondly because part of the kinetic energy is associated with the radial and azimuthal motion. Both these effects contribute terms in r^2 to the velocity which can be neglected for sufficiently small values of r. We may therefore write,

$$\frac{\mathrm{d}}{\mathrm{d}t} = \frac{\mathrm{d}}{\mathrm{d}z} \,.\, v(z),$$

$$\frac{\mathrm{d}^2}{\mathrm{d}t^2} = \frac{\mathrm{d}^2}{\mathrm{d}z^2} \,.\, v^2(z) + \frac{\mathrm{d}}{\mathrm{d}z} \,.\, v(z) \,.\, v'(z), \tag{2.3}$$

where the prime indicates differentiation with respect to z. Equation (2.2) can then be brought into the form,

$$r'' + r'\frac{v'(z)}{v(z)} - \frac{G_1(z)}{mv^2(z)}\, r = 0. \tag{2.4}$$

This is a second order linear differential equation which will have two linearly independent solutions. We will denote these by $r_0 R_1(z)$ and $r_0 R_2(z)$, where r_0 is a constant having the dimension of length and $R_{1,2}$ are non-dimensional functions† of z. The general solution of eq. (2.4) therefore takes the form,

$$\frac{r(z)}{r_0} = c_1 R_1(z) + c_2 R_2(z), \tag{2.5}$$

where c_1 and c_2 are constants to be determined by the boundary

† This notation is at variance with that more generally adopted; the motive is to avoid a notation in which two functions differing only in subscript are dimensionally different.

conditions. We will now choose for our particular solutions two functions which satisfy the following conditions at $z = z_0$:

$$\begin{aligned} R_1(z_0) &= 0 \qquad & R_2(z_0) &= 1 \\ R_1'(z_0) &= 1/r_0 \qquad & R_2'(z_0) &= 0. \end{aligned} \tag{2.6}$$

We will assume that R_1 becomes zero for a second time at $z = z_1$. With this representation, indicated in Fig. 1, eq. (2.5) becomes:

$$\frac{r(z)}{r_0} = r'(z_0) \,.\, R_1(z) + \frac{r(z_0)}{r_0} R_2(z). \tag{2.7}$$

Does eq. (2.7) represent the behaviour of an imaging system? To see that it does, we can write down the form of the equation at $z = z_i$:

$$r(z_i) = r(z_0) \,.\, R_2(z_i). \tag{2.8}$$

Thus the trajectory intercepts this plane at a radius which is directly proportional to the radius at $z = z_0$, and is *independent of the initial slope*, $r'(z_0)$. We can identify the constant of proportionality as the lateral magnification, which we will denote by M:

$$M = R_2(z_i). \tag{2.9}$$

Equations (2.4) and (2.8) are the basis to the study of the behaviour of any axial symmetric force field in the vicinity of the axis. We can now proceed to examine the explicit forms which are relevant to the case when the force field arises from distributions of electric and magnetic fields.

2.2. The ray equation in electric fields

The electric field can be completely described by a potential function in cylindrical polar coordinates $\Phi(r, \theta, z)$. The electric fields which determine the forces exerted on the electron are found by taking the appropriate derivative. The ray equation is derived very directly from the basic equations of motion for the electron, which in cylindrical coordinates and for circular symmetry may be written in the form

$$\ddot{r} = \eta \frac{\partial \Phi}{\partial r}, \tag{2.10}$$

$$\ddot{z} = \eta \frac{\partial \Phi}{\partial z}, \tag{2.11}$$

where η is the ratio of charge to mass, and $\dot{x} \equiv \mathrm{d}x/\mathrm{d}t$.

Equation (2.10) fully describes the radial motion of the electron with which we are primarily concerned. In the above form of the equations, the independent variable is the time. Now, in electron optics it is the variation of the radial distance with position along the axis with which we are primarily concerned. We must therefore utilize eq. (2.11) in order to eliminate time. This is most conveniently accomplished by using the energy integral.

Multiplying (2.10) by $\dot{r}\mathrm{d}t$, (2.11) by $\dot{z}\mathrm{d}t$, adding and integrating, we have

$$\int (\dot{r}\ddot{r}+\dot{z}\ddot{z})\,\mathrm{d}t = \eta \int \left(\frac{\partial \Phi}{\partial r}\mathrm{d}r+\frac{\partial \Phi}{\partial z}\mathrm{d}z\right).$$

Carrying out the integration we obtain the energy integral

$$\dot{r}^2+\dot{z}^2 = 2\eta\Phi+\text{const.} \tag{2.12}$$

Adopting the usual convention, the potential is taken as zero at the cathode. The integration constant is then itself zero, provided that we can neglect the initial thermal velocities. For most purposes this is entirely justified but it is a point to which we will return in our discussion of chromatic aberrations.

Noting that:

$$\frac{\dot{r}}{\dot{z}} = \frac{\mathrm{d}r}{\mathrm{d}z} \equiv r',$$

eq. (2.12) may be written in the form

$$\dot{z}^2 = \frac{2\eta\Phi}{1+r'^2}. \tag{2.13}$$

The second derivative of r with respect to time may be transformed as follows:

$$\ddot{r} = \frac{\mathrm{d}z}{\mathrm{d}t}\cdot\frac{\mathrm{d}}{\mathrm{d}z}\left\{\frac{\mathrm{d}r}{\mathrm{d}z}\cdot\frac{\mathrm{d}z}{\mathrm{d}t}\right\}$$

$$= \dot{z}\left(\dot{z}r''+r'\frac{\mathrm{d}\dot{z}}{\mathrm{d}z}\right)$$

and inserting eq. (2.13), we find the following expression for $\ddot{r}$:

$$\ddot{r} = \frac{2\eta\Phi}{1+r'^2}\left\{\frac{r''}{1+r'^2}+\frac{r'}{2\Phi}\frac{\partial\Phi}{\partial z}+\frac{r'^2}{2\Phi}\frac{\partial\Phi}{\partial r}\right\}. \tag{2.14}$$

We now return to the radial equation of motion, eq. (2.10), and substituting from (2.14) for r find

$$r'' = \frac{1+r'^2}{2\Phi}\left\{\frac{\partial\Phi}{\partial r} - r'\frac{\partial\Phi}{\partial z}\right\}. \tag{2.15}$$

This is the form of the ray equation which we were seeking. It should be emphasized that we have not made any approximations in its derivation; the only essential restriction rests on the assumption of circular symmetry. Most of the work on theoretical electrostatic electron optics is concerned with the analysis and interpretation of this equation. Although formally very simple, the simplicity is obtained at the expense of using the two-dimensional potential function $\Phi(r, z)$ and its partial derivatives. The normal route for exploring the consequences of this equation is to express $\Phi(r, z)$ in terms of the potential distribution $\phi(z)$, *on the axis*. The potential which occurs in the equation is then a function of a single variable (the axial distance z) and we are no longer concerned with a partial differential equation. We shall adopt this course in section 2.4. However, before doing so it is perhaps worth pointing out, that in some cases, where we would like to obtain our solutions by means of a digital computer, the form of eq. (2.15) could be entirely suitable. We shall return to this question in section 3.2.6.

2.3. The ray equation in magnetic fields

Electron motion in the magnetostatic field is inherently more complex than in the purely electrostatic field. The complexity is a direct result of the form of the force term in the Lorentz equation, the component arising from the magnetic field being proportional to $\boldsymbol{v} \times \boldsymbol{B}$. Thus the force depends not only on the local value of the field but also on the local electron velocity. Different electrons passing through a particular point will therefore experience different forces depending on their past histories. Furthermore, the electron will experience forces in the azimuthal direction even though our assumption of circular symmetry excludes the possibility of θ components of the magnetic field. It is convenient to express the magnetic field in terms of the vector potential $\boldsymbol{A}$. As a consequence of circular symmetry only the θ component is non-zero. Since the derivation of the equations for motion in the magnetic field is not complicated in an essential way by the simultaneous presence of electric fields, we will in the following consider the general case in which both are present.

The magnetic field components are related to the vector potential by

$$B_z(r, z) = \frac{1}{r}\frac{\partial}{\partial r}(rA_\theta),$$

$$B_r(r, z) = -\frac{\partial A_\theta}{\partial z}, \tag{2.16}$$

$$B_\theta(r, z) = 0.$$

We may write the force equations in the form

$$\ddot{z} = \eta\left\{\frac{\partial \Phi}{\partial z} - r\dot{\theta}\,\frac{\partial A_\theta}{\partial z}\right\}, \tag{2.17}$$

$$\ddot{r} = \eta\left\{\frac{\partial \Phi}{\partial r} - \dot{\theta}\,\frac{\partial}{\partial r}(rA_\theta)\right\} + r\dot{\theta}^2, \tag{2.18}$$

$$\frac{\mathrm{d}}{\mathrm{d}t}(r^2\dot{\theta}) = \eta\left\{r\dot{z}\,\frac{\partial A_\theta}{\partial z} + \dot{r}\,\frac{\partial (rA_\theta)}{\partial r}\right\}. \tag{2.19}$$

The right-hand side of eq. (2.19) is a perfect differential and the equation may therefore be rewritten in the simpler form

$$\frac{\mathrm{d}}{\mathrm{d}t}(r^2\dot{\theta}) = \eta\,\frac{\mathrm{d}}{\mathrm{d}t}(rA_\theta). \tag{2.20}$$

A single integration then leads to

$$\dot{\theta} = \eta\left(\frac{A_\theta}{r} + \frac{C}{r^2}\right), \tag{2.21}$$

where C is a constant of integration. If we specify $\dot{\theta}$, r and A_θ at the plane $z = 0$ (denoted by θ_0, r_0 and $A_{\theta 0}$),

$$C = \frac{\dot{\theta}_0 r_0^2}{\eta} - A_{\theta 0}\, r_0. \tag{2.22}$$

Eliminating θ in eq. (2.17) and (2.18) by means of (2.21) we obtain

$$= \eta\,\frac{\partial}{\partial r}[\Phi(1 - D^2)], \tag{2.23}$$

$$\ddot{z} = \eta \frac{\partial}{\partial z} [\Phi(1-D^2)], \tag{2.24}$$

where

$$D^2 = \frac{\eta}{2\Phi}\left(A_\theta + \frac{C}{r}\right)^2. \tag{2.25}$$

Equations (2.23) and (2.24) are formally identical to (2.10) and (2.11), the potential function Φ being modified in each case by the factor $(1-D^2)$. By analogy with (2.14), we can at once write down the radial trajectory equation,

$$r'' = \frac{1+r'^2}{2\Phi}\left\{\frac{\partial}{\partial r}[\Phi(1-D^2)] - r'\frac{\partial}{\partial z}[\Phi(1-D^2)]\right\}. \tag{2.26}$$

Equations (2.26) and (2.21) are completely general and suffice to determine the trajectory of an electron in any circular symmetric field. As they stand, in spite of their formal simplicity, they are digestible only by computers. It is possible to transform these partial differential equations into ordinary differential equations, and hence to derive the paraxial forms which are of primary interest, a task which we will now undertake.

2.4. The paraxial ray equation

The approach to a paraxial theory of imaging was outlined in section 2.1. The essential step involves an expansion in powers of the radius r, retaining only the linear terms. The effect of higher order terms are then described as "third order" aberrations, "fifth order" aberrations, etc. This technique for obtaining solutions to the general ray equations is useful, simply because in electron optics, as in light optics, it is precisely the paraxial properties which one generally wishes to exploit. There are indeed cases when the conditions to be met are far from paraxial—for example, wide angle deflexion systems. In such cases, the methods to be described in this section, although they can be pursued to the bitter end, lose all the charms of simplicity.

Our first task is to expand the potential in powers of r. The coefficients in this expansion will be the potential on the axis denoted by $\phi(z)$ and its differentials with respect to z.

If we write

$$\Phi(r,z) = \sum_{n=0}^{\infty} G_n(z) r^n, \tag{2.27}$$

we can determine the coefficients by requiring Φ to satisfy Laplace's equation, which in cylindrical coordinates becomes

$$\frac{\partial^2 \Phi}{\partial z^2} + \frac{1}{r}\frac{\partial}{\partial r}\left(r\frac{\partial \Phi}{\partial r}\right) = 0. \tag{2.28}$$

We can then equate coefficients of like power of r, and find

$$\Phi(r,z) = \phi(z) - \frac{r^2}{4}\phi''(z) + \frac{r^4}{64}\phi'''' \ldots, \tag{2.29}$$

where

$$\phi(z) \equiv \Phi(0,z) \equiv \phi.$$

In a very similar manner we can determine the corresponding expansion for the magnetic field,

$$B_z(r,z) = B(z) - \frac{r^2}{4}B''(z)\ldots \tag{2.30}$$

In the following, we will always understand $B(z)$ to be the axial magnetic field, i.e.

$$B(z) \equiv B_z(0,z) \equiv B.$$

In eq. (2.26), the magnetic field was introduced through the vector potential A_θ. This function can also be expanded, using the first of the eqs. (2.16), in terms of $B(z)$ and its derivatives:

$$A_\theta(r,z) = \frac{r}{2}B(z) - \frac{r^3}{16}B''(z)\ldots \tag{2.31}$$

In the paraxial approximation, it turns out that we need to retain only the first two terms in eq. (2.29) and only the first term in eq. (2.31). In addition we make one further approximation in regarding

$$r'^2 \ll 1. \tag{2.32}$$

For any normal set of trajectories, r' will be, to a first order, proportional to the initial value of r, so that eq. (2.32) is not mathematically a separate condition. However, it is worth noting that the point where the paraxial approximation ceases to apply with the desired accuracy may be set by the magnitude of the slope rather than the magnitude of the radius.

Equation (2.22) shows that for an electron which at any point crosses the axis, or which starts from a source which is outside the magnetic field, the constant $C=0$. We will begin by considering this simpler case. Inserting eq. (2.29) and (2.31) into (2.26), and retaining only the terms linear in r, we find,

$$r''+r'\frac{\phi'}{2\phi}+r\frac{1}{4\phi}\left(\phi''+\frac{\eta}{2}B^2\right)=0. \tag{2.33}$$

Inserting (2.31) into (2.21)

$$\dot{\theta}\equiv\theta'\dot{z}=\frac{\eta}{2}B.$$

With the approximation of eq. (2.32), $\dot{z}\simeq\sqrt{(2\eta\phi)}$,

$$\theta=\sqrt{\left(\frac{\eta}{8\phi}\right)}\int B\,dz. \tag{2.34}$$

Equations (2.33) and (2.34) are perhaps the most important basic equations used in electron optical analysis. It is important to note that the rotation of an electron originating from a point outside the magnetic field, θ, does not, in the paraxial approximation, depend on either the initial radius or its slope. Thus the rotation produced by the magnetic field does not destroy the imaging properties, which, as explained in section 2.1, arise from the presence of a radial force directly proportional to r. Indeed, for many applications, this satisfactory result is all that one wishes to know about the rotation, which otherwise may have little effect on the design of the instrument concerned.

The situation when $C\neq 0$ is not nearly so apparent. However, it has been shown by Bruche by an ingenious method[(7)] that the effect of non-zero C on the paraxial behaviour is solely to add an additional rotational angle to the trajectory, but does *not affect* the radial trajectory equation. Moreover, this additional rotation is, like that of eq. (2.34), independent of initial radius and slope, and therefore does not destroy the Gaussian imaging. This proof, which will not be reproduced here, is important, as it demonstrates for example that small deviations of the source from the axis do not lead to any first order deterioration of the imaging. In the following we will proceed with the simpler case, $C=0$, in the knowledge that this does not affect the radial movement, and hence the focal lengths of lens systems.

In concluding this section we must take note of a transformation of eq. (2.33) which was first suggested by Picht[8] and which has proved to be of great value. If we write

$$R(z)=r(z)\phi^{1/4}(z), \qquad (2.35)$$

eq. (133) becomes

$$R''+TR=0, \qquad (2.36)$$

where

$$T=\left[\frac{3}{16}\left(\frac{\phi'}{\phi}\right)^2+\frac{\eta}{8}\frac{B^2}{\phi}\right].$$

Equation (2.36) is not only significantly simpler than eq. (2.33), but as we will find in later sections, there are other inherent advantages in taking $R(z)$ rather than $r(z)$ as the variable.

2.5. Lens characterization

In eq. (2.33) we have confirmed the conclusion which we had already reached in eq. (2.4) for the general case of a circular symmetric force field—that the paraxial trajectories satisfy a second order linear differential equation. The general solution to any such equation can be expressed as the linear superposition of two particular solutions. A convenient choice of the particular solutions was indicated in Fig. 1, $R_1(z)$ and $R_2(z)$. The initial conditions for these solutions were given in eq. (2.6). On emerging from the force field the trajectories will be linear, and we can define them in terms of their radius and slope at some particular plane located at $z=z_i$. The effect of the force field on paraxial rays can then be completely defined in terms of four constants, $R_1(z_i)$, $r_0R_1(z_i)$, $R_2(z_i)$ and $r_0R_2'(z_i)$. This conclusion holds no matter how complex the lens field, or indeed even when we are considering a combination of a number of distinct lenses.

The way in which these constants are specified and indeed the choice of the two particular solutions is, of course, arbitrary.

For some purposes it is more convenient to define the particular solutions in the more symmetrical manner indicated in Fig. 2. Here $r_1(z)$ is a ray which enters the lens field from the left with zero slope, and $r_2(z)$ is a ray emanating from the same point (the "object"), but starting with an initial slope such that it emerges from the force field parallel to the axis. For a convergent lens these two rays will intersect again at a plane, located at $z = z_i$, a plane which must clearly contain the image. If we now extrapolate the two straight portions of the trajectory $r_1(z)$ they will meet in the plane $z = z_2$, the corresponding intersection for $r_2(z)$ taking place at $z = z_1$. These

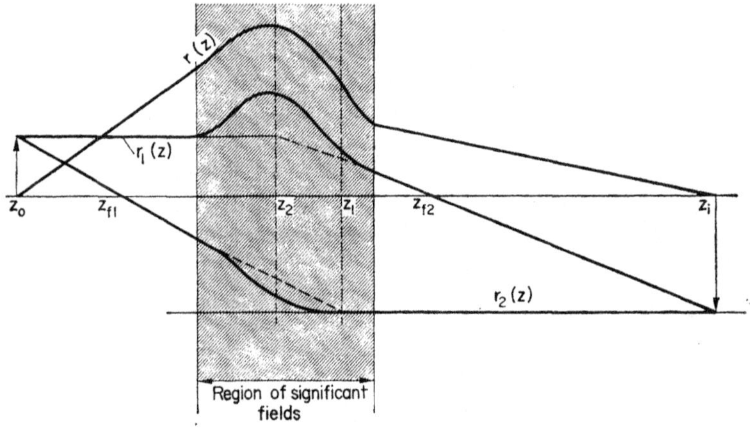

FIG. 2

two planes are called the principal planes of the lens. We can then define two focal lengths in terms of the distance between the intersection with the axis and the principal plane.

Thus, we define†

$$\begin{aligned} f_2 &= z_{f2} - z_2, \\ f_1 &= z_1 - z_{f1}. \end{aligned} \tag{2.37}$$

Armed with the values of f_1 and f_2 and knowing the position of the principal planes, we are able to calculate the position of the image

† The choice of subscripts for the principal planes by which z_2 comes before z_1 is governed by the traditions of classical optics. The principal planes in electron optics invariably appear in the order shown in Fig. 2 in contrast to the usual situation in classical optics. They are therefore frequently said to be "crossed".

for an arbitrary location of an object. Consider a ray $r(z)$ leaving the object at $z = z_0$. We can write $r(z)$ as a linear superposition of r_1 and r_2:

$$r(z) = \alpha r_1(z) + \beta r_2(z), \tag{2.38}$$

where α and β are two dimensionless constants. Consider the particular ray for which $r(z_0) = 0$. The imaging properties of the system which we have quite generally demonstrated in eq. (2.8) then assure us that at $z = z_i$ the ray must cross the axis again so that $r(z_i) = 0$. If we use these two equations to eliminate α and β, we can readily find that (2.38) can be brought into the form

$$\frac{f_1}{u} + \frac{f_2}{v} = 1, \tag{2.39}$$

where

$$u = z_2 - z_0,$$

$$v = z_i - z_1.$$

Equation (2.39) is the thick lens equation familiar from classical optics.

Returning to a consideration of a trajectory such as $r_1(z)$ which enters the lens field parallel to the axis, we are now in a position to determine its magnitude at the point $z = z_i$, and hence to determine the lateral magnification.

On emerging from the field, $r_1'(z)$ is given by

$$r_1'(z) = r_1(z_0)/f_2.$$

Thus

$$r_1(z_i) = [r_1(z_0)/f_2](v - f_2).$$

Thus the lateral magnification M is

$$\frac{r_1(z_i)}{r_1(z_0)} = M = v/f_2 - 1. \tag{2.40}$$

Using eq. (2.39) we can express this result in a more familiar and symmetrical form,

$$M = (v/u)(f_1/f_2). \tag{2.41}$$

For some purposes it is also useful to define an *angular* magnification, M_θ

$$M_\theta = r'(z_0)/r'(z_i).$$

From (2.37), (2.38) and (2.40) one can readily show that

$$M_\theta = v/u. \tag{2.42}$$

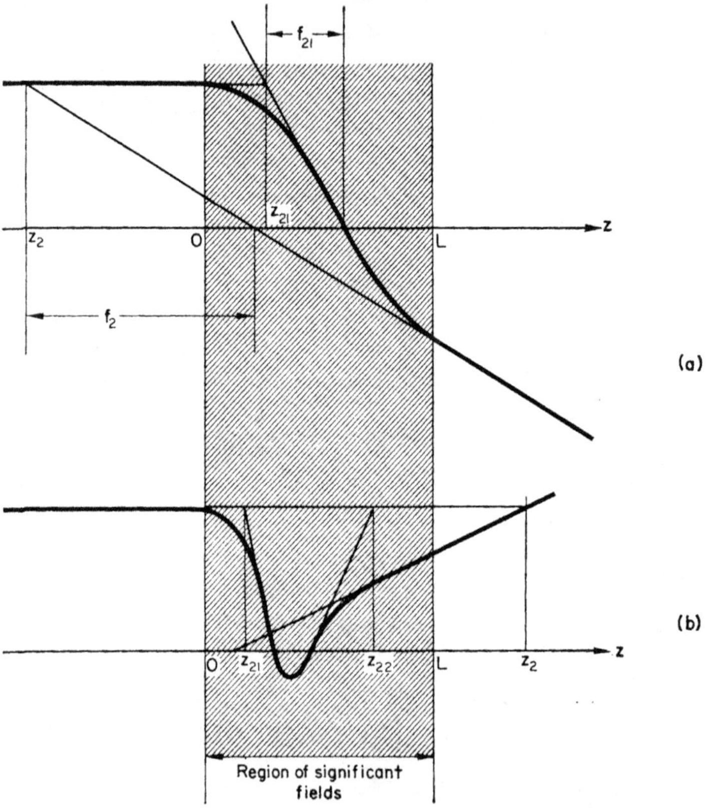

FIG. 3

Thus in a symmetrical lens, the lateral and angular magnifications are equal, whereas in general, for example in the case of a two potential electrostatic lens, they will differ by the ratio of the two principal focal lengths.

Before leaving this topic, we must consider one further complication which can arise in lenses, and which is of considerable practical importance: this is the occurrence of one or more crossovers *within* the lens field. This situation is shown in Fig. 3a for the case of a single crossover. At first sight one might think that no new considerations arise, in that one can still define a focal length

$$f_2 = \frac{r_0}{(r')_{z>L}} \tag{2.43}$$

and a corresponding principal plane. Provided all rays are continuous through the lens such is indeed the case. However, a strong lens of this type is often used as an objective lens, in which the *physical object* is itself placed within the lens field. One is usually concerned with large magnifications, so that the object would normally be very close to one of the crossover points. Under these conditions, it is only the trajectories between the object and the image which is of significance, the trajectories on the other side of the object merely affecting the details of the object illumination. The appropriate focal length which characterizes this use of the lens as an "immersion lens" is clearly

$$f_{2n} = \frac{r_0}{(r')_{z=z_n}}, \tag{2.44}$$

where f_{2n} indicates the focal length associated with the nth crossover point, which we will refer to as the "immersion focal length". This and the corresponding principal plane is also shown in Fig. 3a. Figure 3b indicates a case where there are two crossovers, and where there are consequently two immersion focal lengths in the sense of eq. (2.44) as well as a "normal" one, in the sense of eq. (2.43).

We have now established the framework in terms of which one can express the paraxial behaviour of electron lenses. It should be emphasized that the knowledge of the position of the principal planes and focal lengths suffices for the determination of the paraxial properties only. As we will see in section 4.1, a further eight constants are needed to typify the performance of the lens if the third order aberrations are to be considered.

2.6. Calculation of focal lengths

In general it is not possible to obtain closed form expressions for either the focal lengths or the positions of the principal planes, in terms of the axial potential $\phi(z)$ and the axial magnetic field $B(z)$. There is one notable exception for a magnetic lens, which we will consider in the next section. Fortunately we can make progress by adopting a simplified model—the "thin" lens—which is in many cases a useful approximation and for which the mathematics is much more tractable. For this reason the thin lens theory which is the subject of the following section is much more important to electron optics than in classical optics, where the consideration of a finite thickness does not lead to an essential increase in complexity. The essential simplification is to regard the extent of the refracting field as small compared with the focal lengths.

To calculate the focal length for the general use, we start from eq. (2.36):

$$R''+TR=0.$$

Let the region of appreciable field strength be confined to the range $z_1<z<z_2$. Then we can assume that r' and also R' will be constant for $z<z_1$ and for $z>z_2$. A single integration yields

$$R'(z_2)-R'(z_1)=-\int_{z_1}^{z_2} TR\,\mathrm{d}z. \tag{2.45}$$

The approximation which we are proposing to make is to assume that the variation of R, over the integration range, can be neglected. If this is justified, R may be placed outside the integral. Before the reader too hastily gives his assent to this move, it is only fair to point out that it is not without potential danger. Indeed, it is possible to start from eq. (2.33) without making the transformation to the R variable, and derive an equation similar to (2.45). If we then argue that we can take r outside the integral, we arrive at a focal length which, for the electrostatic lens, is in error by 50 per cent. This erroneous expression is quoted in many references, and it was not until Sturrock[(9)] pointed out the nature of the error that the situation was fully understood. It turns out that using $r(z)$ as the variable, in addition to the contribution to the integral arising from a term quadratic in the field and constant r, there is one arising out of a term linear in the field and linear in r, which must be included.

Returning to eq. (2.45) a detailed inspection of possible higher order terms confirms the validity of regarding R as constant so that

$$R'(z_2)-R'(z_1)\simeq -R\int_{z_1}^{z_2} T\,\mathrm{d}z. \tag{2.46}$$

From eq. (2.43) the focal length can be at once written down:

$$\frac{1}{f_2}=\left(\frac{\phi_1}{\phi_2}\right)^{1/4}\int_{z_1}^{z_2} T\,\mathrm{d}z. \tag{2.47}$$

Here f_2 is the focal length for a ray which is parallel to the axis for $z<z_1$. The complementary focal length for a ray coming from $z>z_2$ is f_1, where

$$\frac{1}{f_1}=\left(\frac{\phi_2}{\phi_1}\right)^{1/4}\int_{z_1}^{z_2} T\,\mathrm{d}z. \tag{2.48}$$

We see that if $\phi_1=\phi_2$ the focal lengths are equal, and that for any thin lens,

$$\frac{f_1}{f_2}=\left(\frac{\phi_1}{\phi_2}\right)^{1/2}. \tag{2.49}$$

Although eq. (2.52), as here derived, is subject to the thin lens approximation, we will see in the following section that it is indeed a rigorous result.

In most cases one in concerned with pure electric lenses ($B=0$) or pure magnetic lenses ($\phi=0$). It is therefore of interest to write the specific form of eq. (2.47) for these two cases:

(i) Electrostatic lens ($B=0$)

$$\frac{1}{f_2}=\frac{3}{16}\left(\frac{\phi_1}{\phi_2}\right)^{1/4}\int_{z_1}^{z_2}\left(\frac{\phi'}{\phi}\right)^2\mathrm{d}z. \tag{2.50}$$

(ii) Magnetic lens ($\phi'=0$, $\phi_1=\phi_2$)

$$\frac{1}{f_2}=\frac{\eta}{8\phi}\int_{z_1}^{z_2} B^2\,\mathrm{d}z. \tag{2.51}$$

In this case we may also be interested in the image rotation, which (for $C = 0$) is simply given by eq. (2.34), the integration limits being z_1 and z_2.

3. DETERMINATION OF FIRST ORDER LENS CHARACTERISTICS

3.1. Analytical solutions

In principle, we can find the focal lengths and the principal planes of a lens by first solving Laplace's equation for the electric and magnetic fields, subject to the boundary conditions imposed by the shape of the lens, and then solving eq. (2.36). It is rarely possible to carry through such a programme both rigorously and usefully; one tends to end up with solutions which in terms of the computation time which they require in their evaluation are at least equally remote from a numerical result as are the starting equations. There are, however, a few simple lenses which can be analysed in an approximate manner and which yield solutions simple enough for slide-rule evaluation. Furthermore, they can give some physical feeling for the behaviour of other, more complicated, lenses. In the following sections we will briefly consider some examples.

3.1.1. THE TWO-APERTURE ELECTROSTATIC LENS

This lens is shown in Fig. 4. Potentials ϕ_1 and ϕ_2 are maintained on the two diaphragms located at $z = 0$ and at $z = L$. If the apertures in the diaphragms are very small compared with their separation, we can assume that the field is confined to the region between the two diaphragms, and is constant.† In this region ϕ'' is therefore zero and, as we shall see, this enables us to integrate eq. (2.33) directly. However, we must also consider the effect of the sudden change in ϕ' which is assumed to take place at each diaphragm. The distribution of the potential which we are assuming and its first two derivatives are also

† Note that it is not correct to take as an idealized model a lens constructed from two planar "electron permeable" conducting membranes. Such a system could be realized (for sufficiently high beam voltages) but it would *not* act as a lens. It is the radial fringing field near the aperture which is responsible for the lens action, a fact which is rather disguised by the use of the constant field approximation.

indicated on Fig. 4, together with the distributions which would actually obtain for a small but finite aperture.

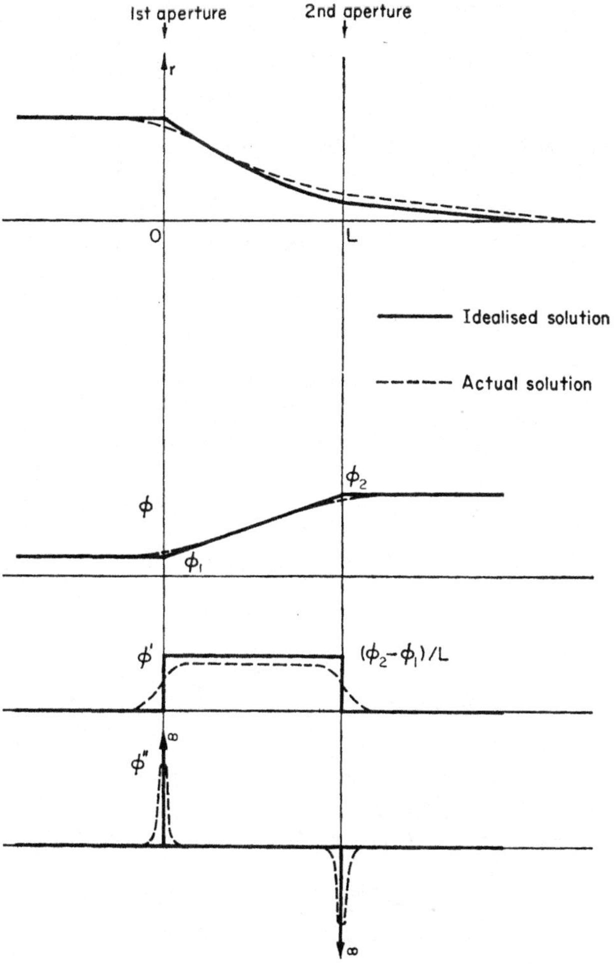

FIG. 4. The two-aperture lens.

We consider an electron with zero initial slope incident at a radius r_0. In the immediate vicinity of the first diaphragm, the dominant influence on the electron arises from ϕ'' which, as seen in Fig. 4, has the form of a delta function. In this (infinitesimal) region we may write eq. (2.33) in the form

$$r''+r\frac{\phi''}{4\phi}=0.$$

Integrating

$$\int r''\,\mathrm{d}z=-\int r\frac{\phi''}{4\phi}\,\mathrm{d}z. \tag{3.1}$$

The integration extends only over an infinitesimal range centred on $z=0$. Since neither r nor ϕ can be discontinuous, they may be regarded as constant in the integration. Noting that immediately to the left of the region r' and ϕ' are zero, we find for small positive values of z

$$\left.\begin{aligned} r &= r_0 \\ r' &= -\frac{\phi'}{4\phi_1}\,r_0 \end{aligned}\right\} \quad \begin{aligned} & z>0 \\ \text{and}\ & \frac{z}{L}\ll 1, \end{aligned} \tag{3.2}$$

where $\phi'=(\phi_2-\phi_1)/L$.

Between the two apertures ($0<z<L$) we assume that ϕ'' is negligible so that

$$r''+r'\frac{\phi'}{2\phi}=0. \tag{3.3}$$

Integrating, we find

$$\phi^{1/2}\,r'=\text{const.}=-\frac{r_0\phi'}{4\phi_1^{1/2}} \tag{3.4}$$

where on the right-hand side we have used the applicable values, found in eq. (3.2). Equation (3.4) may be integrated directly, since within the range $0<z<L$, the potential varies linearly, so that

$$r_L=r_0-\frac{r_0\phi'}{4\phi_1^{1/2}}\int_0^L(\phi_1+\phi'z)^{-1/2}\,\mathrm{d}z$$

and hence,

$$r_L=\frac{r_0}{2}(3-Q^{1/2}), \tag{3.5}$$

where

$$Q=\phi_2/\phi_1.$$

We can obtain the value of r' just to the left of the second aperture from eq. (3.4), i.e. for $0<(1-z/L)\ll 1$:

$$r' = -\frac{r_0}{4}\frac{\phi'}{(\phi_1\phi_2)^{1/2}}. \tag{3.6}$$

When the electron passes through the second aperture it is once again subjected to an impulse, owing to the locally high value of ϕ''. Using the same procedure as in eq. (3.1), we can find that the slope of the trajectory, r'_L, immediately to the right of the aperture, i.e. for $0<(z/L-1)\ll 1$. Since

$$f_2 = -r'_L/r_0,$$

we can immediately obtain an expression for f_2:

$$\frac{f_2}{L} = \frac{-8}{3}(1-Q^{-1})^{-1}(1-Q^{1/2})^{-1}. \tag{3.7}$$

The other focal length is then most conveniently obtained by the use of eq. (2.52):

$$f_1 = f_2 Q^{-1/2}. \tag{3.8}$$

Finally, we can locate the position of the principal planes z_2 and z_1:

$$z_2 = L+(r_0-r_L)/r'_L,$$

$$z_2/L = \frac{1+Q/3}{1-Q} \tag{3.9}$$

and similarly

$$z_1/L = \frac{4}{3(1-Q)}. \tag{3.10}$$

The normalized focal length f_2/L has been plotted in Fig. 5 for values of Q larger as well as smaller than unity. It will be apparent that the corresponding values of f_1 are found simply by using the inverse of Q. As we would expect from the general result obtained in eq. (2.47) f_2 is always positive for a given voltage ratio on the electrodes. The strength of the acceleration lens ($Q>1$) is evidently greater than that for the decelerating lens ($Q<1$). This result applies to all lenses of this form, and is not dependent on the details of the field distribution. The explanation is that the magnitude of the refraction at each aperture is inversely proportional to the local potential. Now one of the refractions is divergent, the other convergent. If $Q>1$, the convergent refraction takes place at the lower of the two potentials and is therefore stronger than when $Q<1$ and it takes place at the higher potential aperture.

When Q exceeds 9, a crossover occurs within the lens field. Although the lens characteristics can still be readily expressed under these conditions, the curves have not been shown beyond this point; one would not generally use such a lens with higher values of Q.

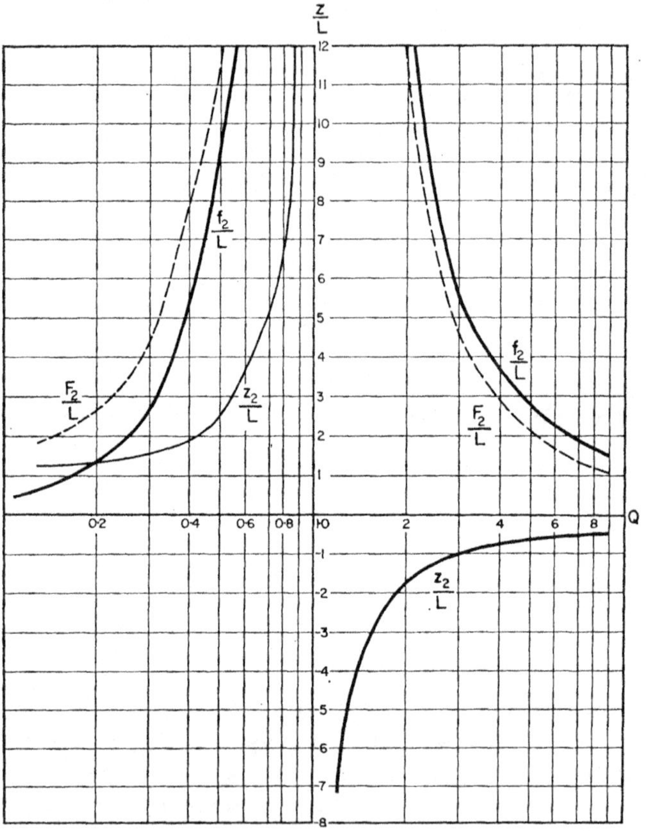

FIG. 5

Figure 5 also shows the location of the z_2 principal plane. It is of interest to note that it never lies within the lens field, and at low lens strengths will be located at large distances from the lens. In practice one might be primarily concerned with the position of the lens focus. We will denote the distance (from the first aperture) to the focus by F $(=f_2+z_2)$ and this quantity is also shown on Fig. 4. We see that for

$Q>1$, F_2 differs from f_2 by an amount which even for $Q=9$ does not exceed 30 per cent. However, for $Q=0.2$, F_2 rises to more than twice the value of f_2.

These observations suggest a question. Under what circumstances can the two-aperture lens be regarded as a "thin" lens. Clearly one condition is that $z_2 \ll f_2$, and, as we can see from Fig. 5, this requirement is satisfied for all values of Q which are not too far from unity. A more pragmatic requirement might be that the thin lens expression for the focal length, eq. (2.47), should give a good approximation to the correct answer. The integral in this equation is readily obtained, and denoting the thin lens focal length by f_{2t}, one can readily show, by comparison with eq. (3.7) that

$$f_2/f_{2t} = \tfrac{1}{2}(Q^{-1/4}+Q^{1/4}). \qquad (3.11)$$

Even for $Q=9$, we see that the thin lens formula is in error by only 15 per cent. We also note that this expression does not change if we invert Q, so that the error for a lens of $Q=5$ is the same as that for a lens with $Q=0.2$. However, there is a pitfall for the unwary in this statement: although the error in f_2 is no worse, the thin lens approximation would also imply that $f_2=F_2$. Thus a simple application of the thin lens formula for the accelerating lens, $Q=5$, would predict the focus with fair accuracy. The focus for the decelerating lens would, however, be in error by a factor of around 2.

Finally, if we wish to describe a lens as thin, it is not enough to demonstrate that for rays going in one direction it behaves as an equivalent thin lens at a particular location—the principal plane. One would also demand that the effective location of the thin lens for rays going in the opposite direction would be in the same vicinity. We would demand in fact that the distance between the principal planes should be small compared with the focal length. From (3.10) we can readily see that for all values of Q, $(z_1-z_2)=L/3$. Thus, in this case, this requirement turns out to be somewhat weaker than that concerned with the relative magnitudes of z_2 and f_2 already discussed; if it is reasonable to regard the lens as thin for electrons going in one direction, it is equally reasonable to regard the equivalent thin lens for electrons going in the opposite direction as being located in approximately the same place.

The case of the two-aperture lens has been discussed in detail, as its behaviour is in many respects typical of other electrostatic lenses. At the same time it is of practical importance, both when used deliberately as a lens, and also when the lens action is incidental to an arrangement of electrodes whose primary function is to accelerate or to decelerate the electrons. Such changes of velocity are, of course, common in a large variety of electron beam equipment.

We will now, however, be in a position to discusss other lenses more briefly and will confine ourselves to sketching the analysis and then presenting results.

3.1.2. THE EINZEL LENS

One frequently wants to achieve a focusing action without causing any ultimate change in the electron velocity. Since, as we have seen in the last section, we obtain a convergent action in a two-aperture lens irrespective of whether we accelerate or decelerate the electrons, we can expect to obtain strong focusing action in the lens indicated in Fig. 6. A particular mode of operation of this lens is to connect the

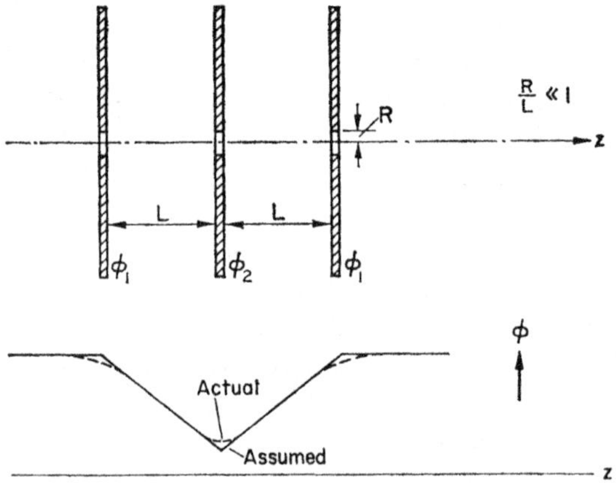

FIG. 6

central diaphragm to cathode potential. With a finite aperture in the diaphragm, the potential on the axis will everywhere remain positive. However, in this arrangement only a single potential supply ϕ_1 is needed. Since in the non-relativistic approximation electron trajectories depend only on the field distribution—not on the absolute magnitude of the field—the focal length of such a lens is independent of ϕ_1 at low velocities, and relatively insensitive to such fluctuations at voltages of the order of magnitude used in electron microscopes. Thus the task of smoothing the power supply for the lens, which is a major consideration in the case of magnetic electron microscope objectives, is very greatly eased.

One can carry out an analysis of such a lens for the case when the aperture diameters are very small as compared with L, in precisely

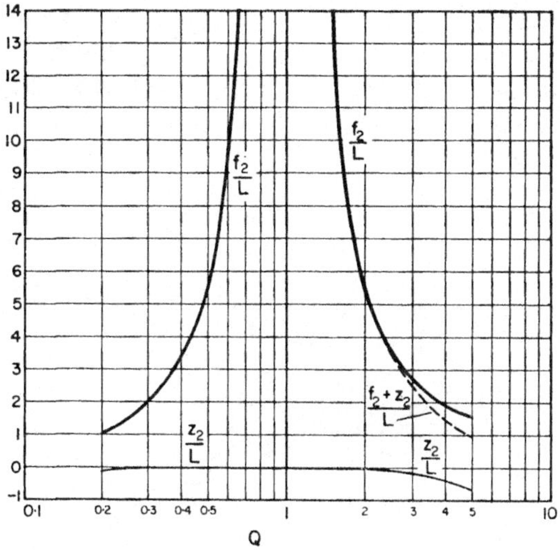

FIG. 7

the same manner as in the previous section. One obtains the following results:

$$\frac{f_2}{L} = -\frac{8}{3}(Q-1)^{-1}(4-3Q^{-\frac{1}{4}}-Q^{\frac{1}{4}})^{-1}, \tag{3.12}$$

$$\frac{z_2}{L} = 1-4(3-Q^{\frac{1}{4}})^{-1}(1+Q^{\frac{1}{4}})^{-1}. \tag{3.13}$$

Since the lens is symmetrical, the other focal length and principal planes are simply given by

$$f_1 = -f_2,$$
$$z_1 = -z_2.$$

The numerical results are shown in Fig. 7; other than the obvious fact that the lens is stronger, the main difference as compared with the two-diaphragm lens is that the principal planes are much closer to the centre of the lens. In particular, for values of Q close to unity they are near the centre of the lens, whereas in the two-diaphragm lens they tend to infinity. In this sense the performance of the Einzel lens is well represented by an equivalent thin lens. However, if we try to calculate the focal length using the thin lens formula of eq. (2.50) we find that the results differ from those obtained from eq. (3.12) quite markedly for $Q>2$; however, for all values of $Q<1$ the approximation is much better[10]. The plot in Fig. 7 has been extended to values of Q where the crossover just occurs within the lens ($Q>4.91$ and $Q<0.2$).

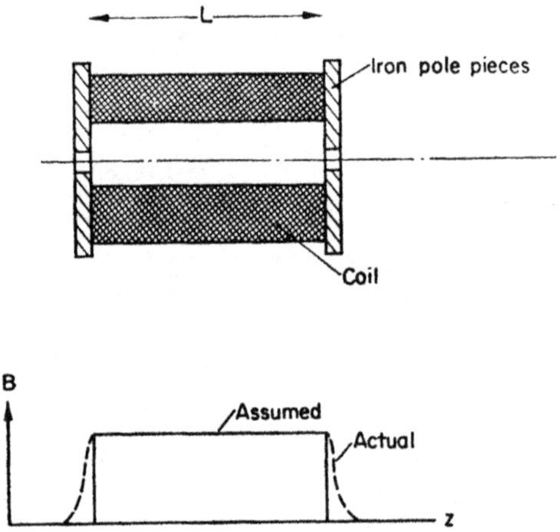

FIG. 8. Uniform field lens.

3.1.3. THE UNIFORM MAGNETIC LENS

The simplest magnetic lens is undoubtedly one where the magnetic field rises suddenly from zero to a constant value which it maintains over a fixed length L. As indicated in Fig. 8 such a lens can be approximated by the use of a solenoid with iron pole pieces. The electric potential throughout is uniform, so that eq. (2.33) becomes,

$$r''+\frac{\eta B^2}{8\phi}r=0. \tag{3.14}$$

Now it is easily demonstrated [from eq. (11)] that an electron injected into a region where the magnetic field has the uniform value B, in a direction at right-angles to that of the field, will describe circles of radius ρ, where

$$\rho^2 = \frac{4\phi}{\eta B^2}. \tag{3.15}$$

Using this radius to express the magnitude of the field† equation (3.14) may be written in the form

$$r'' + \frac{r}{2\rho^2} = 0.$$

For a ray for which $z = 0$, $r = r_0$ and $r' = 0$, the solution is

$$r = r_0 \cos(z/\rho\sqrt{2})$$

and

$$f_2 = -\frac{r_0}{r'_L} = \frac{\sqrt{2}\,\rho}{\sin(L/\rho\sqrt{2})}, \tag{3.16}$$

$$z_2 = L - (\sqrt{2}\,\rho)\,\frac{1-\cos(L/\rho\sqrt{2})}{\sin(L/\rho\sqrt{2})}, \tag{3.17}$$

and from eq. (2.34) we find that the total rotation θ_L of the trajectory around the axis is

$$\theta_L = L/\sqrt{2}\,\rho. \tag{3.18}$$

Notwithstanding their extreme simplicity, these results are not only useful in their own right, but also show many of the characteristics of magnetic lenses with more complex field distribution. We note first that the focal length varies periodically with L, reaching its minimum value for $L = \pi\rho/\sqrt{2}$. For $L = \sqrt{2}\pi\rho$, the focal length is infinite, but, as we see from eq. (3.18), the ray is rotated throughout an angle π; in this case therefore the lens gives an inverted image of an object at minus infinity, at plus infinity.

For short lenses ($L \ll \rho$), eq. (3.16) becomes

$$f_2 = 2\frac{\rho^2}{L} = \frac{8\phi}{\eta B^2 L}. \tag{3.19}$$

† We will use the parameter ρ to describe the effective strength of the magnetic field. In the general case where B is a function of z we will use it to typify the peak value. It is felt that this emphasis on the radius of curvature in a uniform field gives an immediate physical insight into the significance of expressions such as eq. (3.16).

The dependence of the focal length on $1/B^2$ corresponds to that usually observed for any short magnetic lens. Equation (3.19) can also be obtained directly by invoking the thin lens approximation right from the start and using eq. (2.50).

From eq. (3.17) we can see that, again in the limiting case of $L \ll \rho$, $z_2 \simeq L/2$; i.e. that principal plane is located right at the centre of the lens. Even in the extreme case of the lens of maximum strength $(L = \pi\rho/\sqrt{2})$ the principal plane has moved only 0.14 L from the centre of the lens.

3.1.4. THE BELL-SHAPED MAGNETIC LENS

The lenses which we have so far discussed in this section are all at least one step away from reality. Almost invariably the use of a field distribution which more closely approximates real lens situations leads to analytical expressions which, in the sense discussed in section 1, are not "solutions"—they are as far from numerical results as the starting equations, and they reveal nothing by inspection. A notable exception is the bell-shaped field which has been investigated in great detail by Glaser[(11)]:

$$B = \frac{B_0}{1+(z/a)^2}. \tag{3.20}$$

This field distribution is reasonably similar† to that produced by any lens with a fairly narrow gap between the iron pole pieces—a design invariably adopted in electron microscope and cathode ray tube lenses. Substituting the distribution into eq. (2.33) leads to a non-linear equation which Glaser, by means of two ingenious transformations, was able to reduce to a simple linear form.

Lenses of this type are often used as immersion lenses, i.e. with the object to be imaged located within the lens field. As discussed in section 2.5 when an object is situated close to a crossover within the lens field so that it is imaged with large magnification, it is useful to define an "immersion focal length" which differs from the normal focal length for uninterrupted trajectories. Both are of interest; the former when the lens is used as an immersion objective, the latter when the object is outside the lens field or also whenever the "object"

† In fact it decays somewhat too slowly with distance from the centre.

is virtual, as occurs for example for the intermediate and projection lenses in an electron microscope.

The result for the "normal" focal length and principal plane takes the form

$$f_2 = \frac{r_0}{(r')_{z \gg a}} = \frac{a[1+\frac{1}{2}(a/\rho)^2]^{1/2}}{\sin \pi [1+\frac{1}{2}(a/\rho)^2]^{1/2}}, \tag{3.21}$$

$$z_2 = a\left[1+\tfrac{1}{2}\left(\frac{a}{\rho}\right)^2\right]^{1/2} \cot \frac{\pi}{2}\left[1+\tfrac{1}{2}\left(\frac{a}{\rho}\right)^2\right]^{1/2}. \tag{3.22}$$

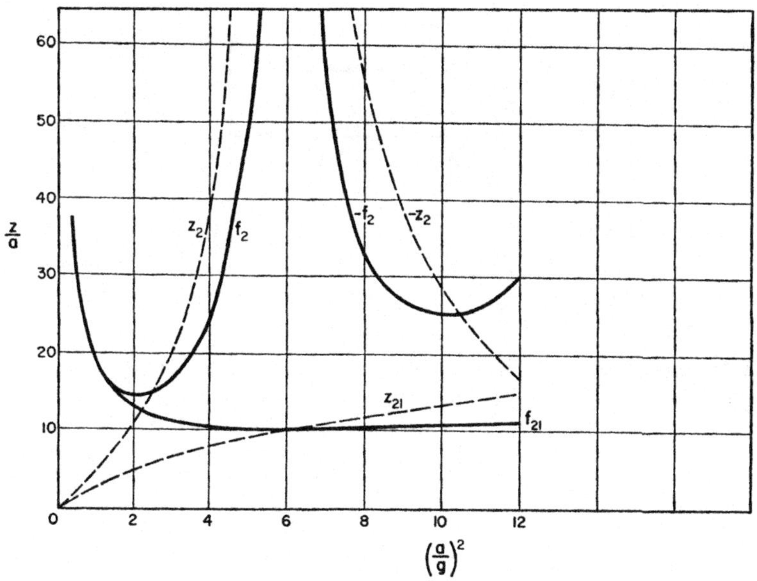

FIG. 9. Focal characteristics of the Glaser lens.

Numerical results are plotted as a function of $(a/\rho)^2$ in Fig. 9. We see that as in the case of the uniform magnetic lens field, the focal lengths change sign periodically, although, in the present case, the focal lengths increase in each successive period. Each change of sign in f_2 corresponds to an additional crossover within the lens field. As is readily seen from eq. (3.21) the total number of crossovers is equal to N, where N is the largest integer for which

$$N < \left[1+\tfrac{1}{2}\left(\frac{a}{\rho}\right)^2\right]^{1/2}. \tag{3.23}$$

Thus for $\frac{1}{2}(a/\rho)^2 < 3$ there is only one crossover, for $\frac{1}{2}(a/\rho)^2 < 8$ there are two, etc. In each period there is a well-defined minimum which, however, increases progressively as we go to higher values of N. The form of the trajectory for $N = 1$ and $N = 2$ is indicated in Fig. 10.

The "immersion focal length" and corresponding principal planes are defined in terms of the slope of the trajectory at the nth crossover

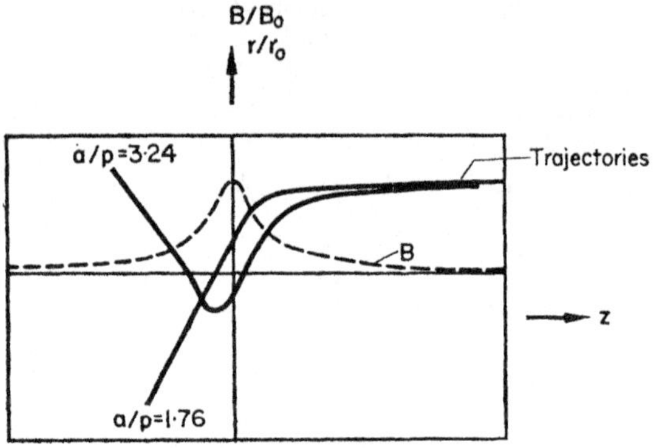

FIG. 10. Trajectories in the Glaser bell-shaped field.

point which we will define as occurring at $z = z_n$; naturally n must be less than N.

$$f_{2n} = \frac{r_0}{(r')_{z = z_n}}$$

$$= -\frac{(-1)^n a}{\sin n\pi\,[1+\frac{1}{2}(a/\rho)^2]^{-1/2}}, \tag{3.24}$$

$$z_{2n} = -(-1)^n a \cot\frac{n\pi}{2}\left[1+\frac{1}{2}\left(\frac{a}{\rho}\right)^2\right]^{-1/2}. \tag{3.25}$$

It is readily seen that here, once n is fixed, the sign of f_{2n} and z_{2n} does not vary with the parameter a/ρ. In practice one is almost invariably concerned with the case when there is only one crossover—although this may be well inside the lens field. The values of f_{21} and z_{21} have been plotted in Fig. 9. We note that for values of $\frac{1}{2}(a/\rho)^2$ less than around 2, there is very little difference between f_2 and the "immersion focal length" f_{21}. This is, of course, to be expected because when the focal length is large, the crossover will occur at a point where the field strength is already very low, the gradient of the trajectory r' is accordingly very close to its ultimate value reached at points far from the lens. The main differences for larger values of the abscissa are that the focal length reaches a very flat minimum at $(a/\rho)^2 = 6$—at the very point where the normal focal length becomes infinite. This is readily understood when it is appreciated that this condition is just that where a second crossover point is formed. The flatness of the minimum shows that one can realize a lens, having very nearly the maximum strength, without approaching this point too closely. For example, at $(a/\rho)^2 = 3$—one-half of the saturation value—the focal length is only 10 per cent above the minimum value.

3.2. Numerical and experimental determination of lens characteristics

If the focal lengths of a lens are large as compared with the extent of the field (i.e. a "weak" lens), one would expect, and our calculations in the previous section confirm, that the thin lens expression of eq. (2.50) will be a good approximation. It is often possible to represent an experimentally determined field distribution with good accuracy by an analytical expression, which allows one to integrate eq. (2.50). If, on the other hand, the lens is strong, invariably the case in electron microscopes and related apparatus, purely analytical methods based on the general expression of eq. (2.48) are not likely to succeed.

In the last section we have discussed a few simple cases where useful analytical results emerged. This list of cases is not exhaustive. Amongst other solutions which have been obtained we will note one for the two-cylinder electrostatic lens[12], and solutions for the magnetic lens with a $1/\cosh(kz)$ distribution of the field[13,14]. The latter complements the bell-shaped distribution discussed in the previous section in that it is most nearly applicable to a very narrow gap lens, whilst the bell-shaped distribution corresponds more closely to the wide gap case. Solutions for a number of variants of the basic bell-shaped distribution have also been reported[15,16]. Finally mention should be made of an analytical solution of the electrostatic Einzel by Glaser and Schiske[17] later extended by Archard[18].

Until recently, there were two main sources of information on the focal properties of practical lenses: direct experimental data and data derived from analogue or hybrid analogue/digital techniques.

Whilst most of the data on lenses currently available was obtained in one of these ways, with present-day computers, we must now also consider direct digital computer evaluation, in which *both* the field distribution and the trajectories are computed digitally. It is the author's view that the calculation of the characteristics, both first and third order, of a new (circular symmetric) lens is most conveniently carried out in this way. For this particular purpose the digital computer supersedes analogue techniques—in the sense that the cost involved in getting an answer will be less.

It will not be possible, within the scope of this work, to discuss either experimental or numerical methods in any detail. In the following sections we will therefore confine ourselves to some brief comments, and references to the literature for the interested reader.

3.2.1. EXPERIMENTAL DETERMINATION OF FOCAL PROPERTIES

The basic method of measurement is to obtain the location of object and image distances. One such measurement suffices to determine the focal length if the position of the principal plane can be assumed to be known (e.g. as in the case of a weak symmetrical Einzel lens). To determine the position of the principal plane as well, one can either repeat the measurement for a different object distance, or alternatively seek to measure the angle of incidence of electrons arriving at the image plane at a known radius.

A large variety of similar experimental arrangements have been used. Figure 11a shows the basis of a method pioneered by Klemperer[19], which is particularly suitable for relatively weak lenses, such as used in cathode ray tubes and similar apparatus. The gun produces a parallel beam of electrons, and a sample is allowed to pass through the small aperture whose distance from the axis can be varied. The fluorescent screen can be moved in an axial direction and its position adjusted so that the point of impact is on the axis, observation being effected by means of a microscope. Observation of the radial movement of the spot as the screen is moved along the axis then determines the angle of the ray. This information suffices for the determination both of focal length and of the principal plane.

An advantage of this technique is that it also enables one to explore aberrations. For example, by noting the change in effective focal length with the radial distance of the aperture, one can obtain a value for the spherical aberration constant (see section 4). Other aberrations can also be explored in this way.

The need for movable screens and apertures can be eliminated by the use of a pair of grids placed before and after the lens, using essentially point source illumination. This method, due to Spangenberg and Field[20], was used to obtain a large body of data on simple

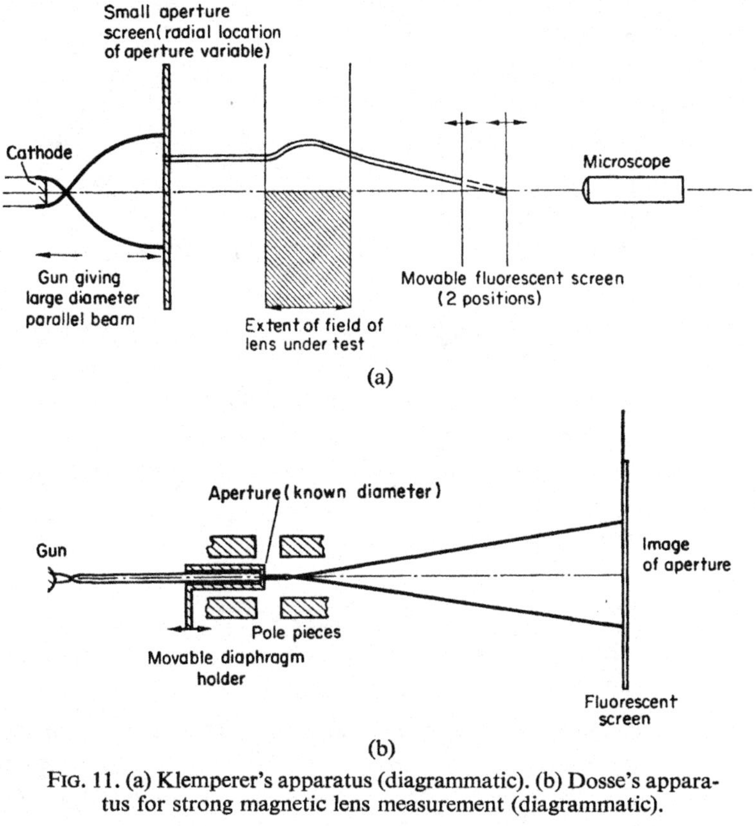

FIG. 11. (a) Klemperer's apparatus (diagrammatic). (b) Dosse's apparatus for strong magnetic lens measurement (diagrammatic).

electrostatic lenses, one example of which is presented in Fig. 12. However, with modern vacuum techniques, simple and reliable movable seals are readily available, and the method shown in Fig. 11a would appear to be simpler and more direct.

This method is not applicable to a high-power lens such as a magnetic electron microscope objective, simply because the focal plane will in general be within the lens field, and in any event impossible to locate by means of a movable screen.

One approach to this situation[21] is indicated in Fig. 11b. Here a small aperture of known diameter is located on the axis, and a strongly magnified image is formed on the screen. In the case of a magnetic objective, the aperture can be moved right through the lens field, and by measuring magnification and image distance for several object distances both the focal length and the position of the principal plane can be found.

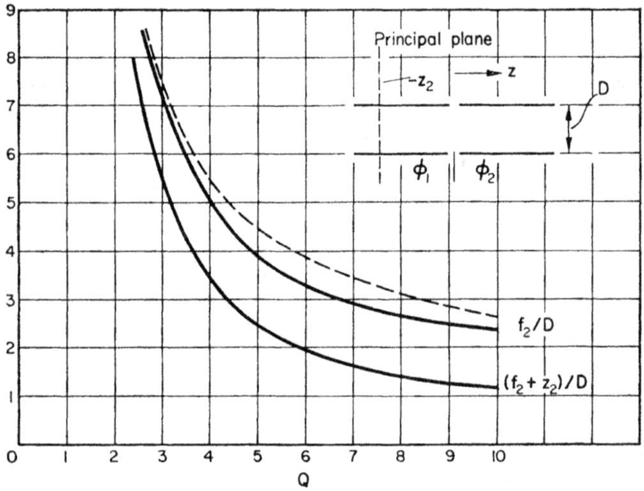

FIG. 12. Focal properties of two-cylinder accelerating lens. Comparison between Spangenberg and Field experimental results (solid line) and ray integration for approximate field distribution by Zworykin *et al.* (dotted). Discrepancy for $(f_2+z_2)/D$ is negligible.

For short focal length electrostatic lenses one can measure the lens[22] characteristic by observing the shadow cast by a small aperture placed in turn on either side of the lens, and illuminated by a small source.

3.2.2. EXPERIMENTAL DETERMINATION OF FIELD DISTRIBUTION

It is not possible, at least by any simple means, to determine the electric field distribution in an electric lens by direct measurement. There are, however, several methods available for measuring the magnetic field distribution in a magnetic lens. If the lens is sufficiently large, the simplest method is to place a search coil on the axis and use a ballistic technique. An elegant variant on this basic scheme due to Fert and Gautier[23] is indicated in Fig. 13. A coil is located on the axis of the lens and is vibrated. The coil is sufficiently long to ensure that the field strength at one end is negligible. The a.c. output is then proportional to the field strength at the other end. By moving the mean position of the coil, the complete field distribution can be derived. In order to determine the paraxial characteristics, a knowledge of $B(z)$ is sufficient [eq. (2.50)]. However, in order to compute the aberration coefficients a knowledge of $B'(z)$ and $B''(z)$ is also needed. The former of these quantities is measured by means of a very thin coil, C_1, the output naturally being proportional to $B'(z)$

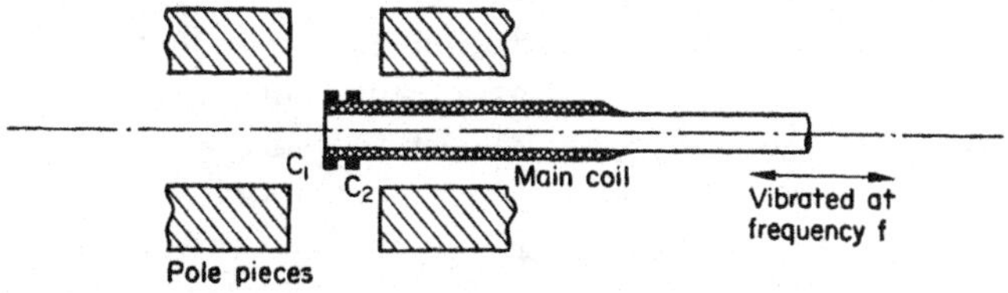

FIG. 13. Determination of distribution of $B(z)$, $B'(z)$ and $B''(z)$.

(no output would be obtained in a uniform field.) The second differential is obtained by utilizing two coils, C_1 and C_2, spaced by a small distance, which are carefully wound to be as similar as possible, and connected in opposition. This technique has achieved accuracies in the field distribution which are of the order of plus or minus 0.1 per cent of the peak field.

It is difficult to make search coils with diameters much under 0.5 mm. Since this dimension is typical of the diameter of the bore of an electron microscope objective lens one cannot hope to obtain a very detailed picture of the distribution in this way. However, by careful design of the search coil[24, 25] parameters it is possible to reduce first order errors arising from the finite size of the coil.

An alternative technique is to measure the field by means of the Hall effect. So far Hall probes with an active area of less than 2 mm do not appear to be commercially available—although one would think that using modern semiconductor device technology much smaller devices could be fabricated. The technique already described for measurement of the first and second derivatives of the magnetic field distribution is, of course, also applicable to the Hall technique.

3.2.3. ANALOGUE METHODS FOR FIELD DISTRIBUTION MEASUREMENT

Both the electrostatic potential and the scalar magnetic potential satisfy Laplace's equation. Thus any system which is also subject to Laplace's equation can, in principle, be used as an analogue for the problem of field distribution. The most commonly used analogue is based on the fact that the distribution of potential in a homogeneously conducting medium satisfied Laplace's equation. The medium can take on a variety of physical forms, including the "electrolytic tank" or resistance paper for systems of appropriate symmetry. A very important variant on these schemes is the use of networks consisting of large numbers of separate resistance elements, an apparatus which could be regarded as a "finite difference", or discrete, analogue of a conducting medium. Another analogue is represented by the displacement of a stretched elastic membrane, a model which was used in the hey-day of the grid controlled valve, and is still of some value in the design of such complex, but relatively low precision, devices as electron multipliers. It is an analogue which allows direct representation of the electron trajectories, but is less adaptable to the separate measurement of field distribution.

A very great amount of effort has, in the past, been devoted to the perfection of analogue techniques; within the confines of this survey we can only list a few references to the extensive literature. It is a subject of considerable importance, in that our most detailed knowledge of the characteristics of electron microscope objectives is derived from analogue (resistance network) techniques. At the same time it is felt that these techniques are, for most applications, obsolescent and it is probable that they will be very largely replaced by purely digital methods.

The electrolytic tank is certainly the most widely used analogue, and has a history which goes back to the last century[(26)]. In principle it consists of a tank filled with a fairly low conductivity liquid—in some places the tap water is about right. The electrodes of an electrostatic lens are simulated by suitable formed metal electrodes. In most cases one is concerned with circular symmetry. In this case it is usual to use a tilted tank so as to form a fairly thin sector-shaped conducting medium, the thin edge of the wedge corresponding to the axis of symmetry. However, it has been found that for greater accuracy it may be preferable to use a complete half-section of the model bounded by an insulating plane. The potential distribution is then obtained by means of probe measurements. In some cases the field rather than the potential is directly measured by means of a double probe with a differential voltmeter. The method is simple in principle, but a large number of difficulties have to be overcome in order to achieve accurate results. In practice, accuracies of around 1 per cent of the maximum field are fairly readily achieved. With great care the errors can be reduced to 0.1 per cent. In the case of magnetic fields the procedure is very similar, in this case metal electrodes representing the iron pole pieces. This representation assumes that the permeability is sufficiently large to be regarded as infinite. In practice the finite permeability of iron may be of significance, particularly under conditions when the flux densities approach saturation values. In principle one could take such effects into account in an electrolytic tank by the use of a high conductivity liquids to represent the pole pieces. However, in practice it is very difficult to implement this idea, since in cases of practical importance the effective permeability will vary over the pole piece.

A detailed description of this technique, and references to work up to 1950, will be found in a review article by Liebmann[(27)]. Much of the emphasis since then has been on the extension of the technique to include space charge effect; although these are not directly relevant to our present subject, the technological improvements which have been made have led to a substantial improvement in the overall accuracies which can be achieved[(28, 29, 30)].

The resistance network functions in a way which is essentially identical to the operation of the electrolytic tank. It starts off with the not insignificant advantage that it is not wet. The accuracy depends ultimately on the mesh size, i.e. on the coarseness of the net which is to simulate a uniform medium. However, it turns out that one can greatly increase the overall accuracy by using a very fine mesh in the regions where the field variations are particularly rapid, and a much coarser mesh elsewhere. The accuracy also, of course, depends on that of the individual resistors and the technique for potential measurement at the mesh points. An accuracy of around ± 0.01 per cent of the maximum field has been obtained[(31, 32)].

The relative advantages of the tank and the network for field distribution measurements have been hotly, though somewhat inconclusively, contested. The author, who has used tanks but not resistance networks, is inclined to side with the adherents of the latter.

3.2.4. DIGITAL COMPUTER METHODS FOR FIELD DISTRIBUTION

The essence of all numerical procedures for the solution of a boundary value problem is to replace the actual equation by its finite difference approximate form. The evaluation of the field is confined

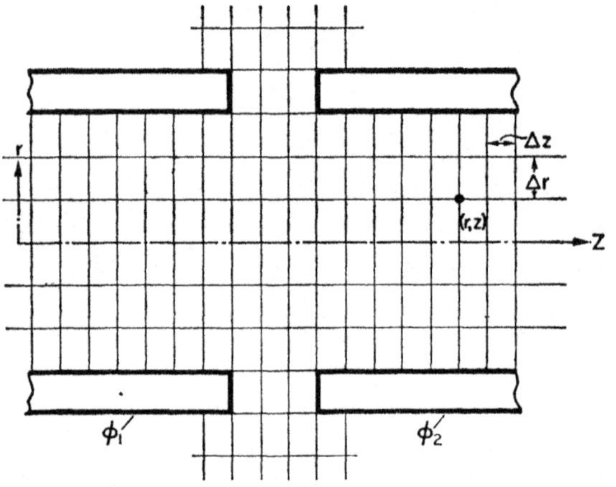

FIG. 14. Mesh for finite difference solution of two-cylinder potential problems.

to the mesh points of a grid which covers the whole region. Figure 14 shows such a mesh arranged for the solution of a two-cylinder electrostatic lens. The Laplace equation for the potential is

$$\frac{\partial^2 \Phi}{\partial r^2}+\frac{1}{r}\frac{\partial \Phi}{\partial r}+\frac{\partial^2 \Phi}{\partial z^2}=0. \tag{3.26}$$

If we now consider a typical mesh point (r, z), we can write down the usual approximate expressions for the required derivatives in terms of the potentials at the adjacent mesh points, and the mesh intervals in the z and r direction, Δ_z and Δ_r:

$$\frac{\partial\Phi}{\partial r}=\frac{1}{2\Delta_r}[\Phi(r+\Delta_r,z)-\Phi(r-\Delta_r,z)],$$
$$\frac{\partial^2\Phi}{\partial r^2}=\frac{1}{\Delta_r{}^2}[\Phi(r+\Delta_r,z)+\Phi(r-\Delta_r,z)-2\Phi(r,z)],$$
$$\frac{\partial^2\Phi}{\partial z}=\frac{1}{\Delta_z{}^2}[\Phi(r,z+\Delta_z)+\Phi(r,z-\Delta_z)-2\Phi(r,z)].$$

To this approximation Laplace's equation becomes

$$\left(1-\frac{\Delta_r}{2r}\right)\Phi(r+\Delta_r,z)+\left(1-\frac{\Delta_r}{2r}\right)\Phi(r-\Delta_r,z)+\left(\frac{\Delta_r}{\Delta_z}\right)^2[\Phi(r,z+\Delta_z)$$
$$+\Phi(r,z-\Delta_z)-2\Phi(r,z)]-2\Phi(r,z)\left[1+\left(\frac{\Delta_r}{\Delta_z}\right)^2\right]=0. \quad (3.27)$$

This means that once a solution has been found this equation will be satisfied to an accuracy which depends on the fineness of our mesh.

The first step in the procedure for finding a solution is to make a guess at the potential distribution, and then by the use of eq. (3.27) to evaluate the "error" at each mesh point. The errors are then reduced by modifying each mesh point value, and simultaneously the four adjacent mesh points according to a definite scheme which eventually enables one to reduce all the errors below any specified value. This procedure was first applied to electron optical problems by Liebmann[(27)]; it is a method which is now seen to be a particular example of a very general class of numerical techniques which are known as "relaxation methods"[(33, 34)].

It is a procedure which is well adapted to computer calculation; indeed it is interesting to note that as early as 1943 Kormes[(35)] published a paper with the title "Numerical solution of the boundary value problem for the potential equation by means of punched cards".†

The basic method outlined above has developed into a sophisticated branch of numerical analysis; procedures for accelerating the convergence of the process[(36, 45)] and for estimating the resulting error have been developed, as well as criteria for the optimum distribution of mesh sizes in different parts of the whole region. Once a

† The author notes that the calculations were carried out by "ordinary machine operators working from written instructions and without supervision". Also "... that the method goes beyond the particular problem treated in this paper and that it can be successfully applied to other problems involving successive approximations".

basic programme has been developed solutions to new problems are obtained rapidly, with any desired accuracy, and economically. The development of such a programme, and more particularly one that also gives trajectories, is, of course, a major task, but it has to be compared with the task of building up and analogue device such as an electrolytic tank. Furthermore, programmes are *transferable* and once completed one can be used simultaneously by a large number of groups in different parts of the world.

3.2.5. ANALOGUE METHODS FOR TRAJECTORY DETERMINATION

It was shown by Gabor[37] in 1937 and independently by Langmuir[38] that the concept of the electrolytic tank could be extended to trajectory determination, and moreover that this extension can be made automatic. It can be shown that the trajectory equation can be expressed in the form

$$R_c = \frac{2\Phi}{\Phi'_n}, \tag{3.28}$$

where R_c is the instantaneous radius of curvature of an electron at the potential Φ subjected to a field Φ'_n normal to the direction of motion. In Gabor's calculator, a system of probes measured Φ and Φ_n by means of a special bridge and by the use of a mechanical linkage system; this information was used to drive a trolley, connected to the probe system, along the electron trajectory. This method was later extended by a number of workers; the most successful embodiment is a hybrid version, where the probe information is transformed by digital techniques to predict the required probe movement[29, 30].

An entirely different analogue trajectory tracer, and undisputably the simplest of all, is the elastic membrane, used in conjunction with a camera, to record the motion of a small ball bearing over its surface. If the surface of a stretched elastic membrane is perturbed at one point, the displacement at other points is a solution of Laplace's equation. If the surface is subjected to displacements at places corresponding to the position of electrodes, and to a height which corresponds to their potential, the displacement at all points will represent the potential distribution. This analogue can naturally only apply to *planar* fields. It is therefore suitable for studying deflexion systems, gridded tubes, electron multipliers, etc., but is not adaptable to rotationally symmetric situations. The motion of a small ball-bearing on the surface of the displaced field can be shown to correspond to the motion of an electron in the electrostatic field[39].

The analogue is approximate in that it neglects the rotational energy in the ball-bearing, and also the effects of friction. However, Walker[40] has shown that if sufficient care is taken, quite surprisingly high accuracies can be achieved. The model is of particular value in the preliminary design of electron multipliers[41], and of deflexion systems. The aim here is a rather simple one—to ensure that the electrons hit rather than miss the next dynode. For this purpose the accuracy of the analogue is adequate.

3.2.6. DIGITAL METHODS FOR TRAJECTORY DETERMINATION

The distribution of the field must naturally be the starting point for any trajectory computation. In practice this might be derived from one of a number of possible sources:

(i) an analytical solution of the potential problem;
(ii) a digital computer solution of the potential problem;
(iii) experimental or analogue data, usually in the form of the potential distribution on the axis.

In the case of either (i) or (ii), the complete distribution is available, and we are therefore in a position to effect a finite difference solution of eq. (2.26) which, with the given data, simply takes the form

$$r'' = (1+r'^2).f_1(r,z)+r'(1+r'^2).f_2(r,z), \tag{3.29}$$

where f_1 and f_2 are known functions of r and z, which are directly computed from the basic field distribution data. The numerical solution of this equation has been discussed by a number of writers. An approach, by Liebmann[42] was used by him to obtain accurate first and third order characteristics of a magnetic lens. The essential step is to expand the radius r_{n+1} and the slope r'_n, the values at the end of the nth interval, in a Taylors series

$$\begin{aligned} r_{n+1} &= r_n + r'_n\,\Delta z + r''_n\left(\frac{\Delta z}{2}\right)^2, \\ r'_{n+1} &= r'_n + r''_n\Delta z + r'''_n\left(\frac{\Delta z}{2}\right)^2 \end{aligned} \tag{3.30}$$

the coefficients, obtained directly from eq. (3.29), taking on the mean value in each interval Δz. A development of this basic method, which gives a more rapid convergence has been described by Burfoot[(43)].

The advent of the high-speed computer has not only enormously facilitated the business of getting numerical results[(29,44,45,109)]—it has also relieved the toiler in electron optical fields from the necessity of knowing very much about numerical methods. He will find that subroutines for the solution of equations such as (3.29) are part of the normal "software" of most computers. Given a little luck, the user can safely remain in ignorance of the finer points of the working of the routine.

In the case of (iii), when the field data are known only on the symmetry axis, one can use the paraxial equation (2.36) if only the Gaussian characteristics of the lens are required. If on the other hand it is desired to explore the third order aberrations as well, a somewhat more involved equation, which includes higher derivatives of the potentials, must be used. This equation is given in section 4.21, eq. (4.7). The programming of this equation again presents no difficulties. In either case, given the field distribution in some form, it is, in principle, a relatively simple matter to obtain the trajectories.

As one example of a computation of this type, the trajectories through an "Einzel" lens were calculated for an axial distribution of the potential given by

$$\phi = \frac{\phi_0}{1+\alpha}\left(1+\alpha\cos\frac{2\pi z}{L}\right). \tag{3.31}$$

The ratio of the potential at the centre of the lens to the value at either side is therefore Q, where

$$Q = \frac{1-\alpha}{1+\alpha}.$$

Results,† for two values of Q are shown in Fig. 15. In each case the plot shows the paraxial trajectory calculated from eq. (4.7) (for rather extreme situations where the initial radius is 0.5 L, the half-length of the lens). The trajectories, accurate to better than 1 part in 10^4 of the initial radius are computed on a high-speed computer (Atlas) in about 0.5 sec.

Anticipating the discussion of section 4, one should add that often

† These data were kindly supplied by Mr. O. Penge.

the complete trajectory is not the information which is most directly useful. One is normally concerned with the Gaussian characteristics, and with an evaluation of a specific type of aberration. The method by which differences between the paraxial and the true trajectories can be analysed into specified aberrations will be discussed in section 4.1.

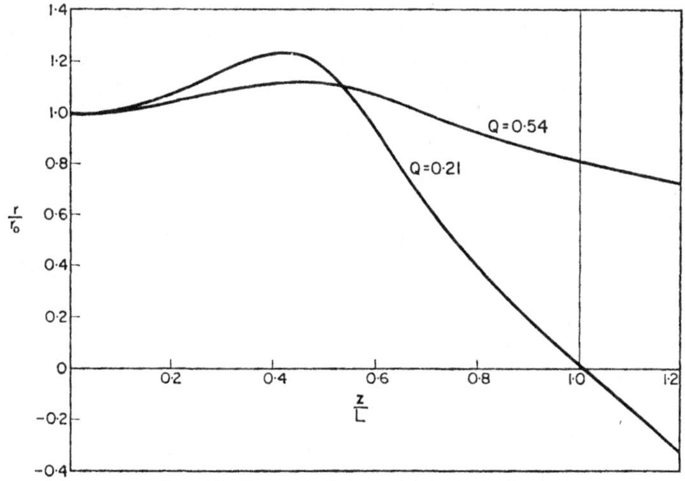

FIG. 15. Computed paraxial trajectories.

3.3. Lens characteristics—results

In this section we will present some data relating to practical lenses derived either from experimental results or from the numerical development of reliable analogue data. Of the great number of results which have accumulated, we will have space to note only those relating to the most frequently used lens arrangements. However, the location of some further information, which is very widely scattered in the literature, will be noted.

A very widely used electrostatic lens consists of two closely spaced equidiameter cylinders. This type of lens is frequently used in low-current electron guns. In addition it is often important to know the strength of such a lens arising "accidentally"—i.e. whenever one accelerates or decelerates an electron beam by the use of a two-cylinder structure. This case is one of the few for which a reasonably accurate analytical result for the axial potential distribution is available. The results of a numerical integration of the ray equation[47] are compared with experimental results in Fig. 12.

An extension of the theory, for cylinder lenses of unequal diameter, has been carried out by Ollendorf[48]. The results must be regarded as somewhat less reliable than those for the equidiameter lens, but adequate for many purposes. The result, for the particular case when the ratio of the two-cylinder diameters is 2:1, is given in Fig. 16.

Detailed experimental results relating to two cylinder lenses of equal and unequal diameters, and including the effect of a finite spacing between the two elements are presented in Spangenberg's

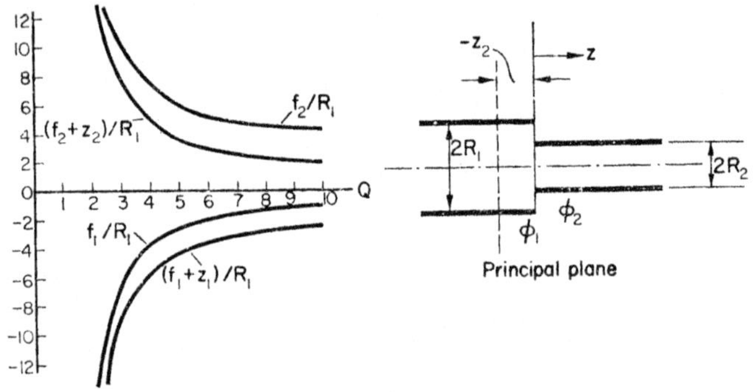

FIG. 16. Focal properties of a two-cylinder lens, $R_2/R_1 = 0.5$ (after Ollendorf).

book[49]. This collection of results also includes data on lenses formed by one cylinder and one apertured diaphragm.

For reasons which were briefly noted in section 3.1.2, the three-diaphragm lens, with the outside diaphragms connected to the same potential, is particularly useful. Such lenses are often referred to as "Einzel lenses", a term which strictly speaking is appropriate only for the case when the central electrode is at the cathode potential, as would invariably be the case when the lens is used as the objective in a microscope. However, in an electrostatic microscope it is necessary to vary the magnification over wide limits, so that the focal length of the intermediate lens and possibly also of the projector lens must be adjustable. One is therefore in practice concerned with the characteristics of such lenses for values of the central potential covering a small range about zero (i.e. cathode potential).

Data for such lenses, based on an electrolytic tank evaluation of the field distribution, were first given by Ramberg[50]. More detailed

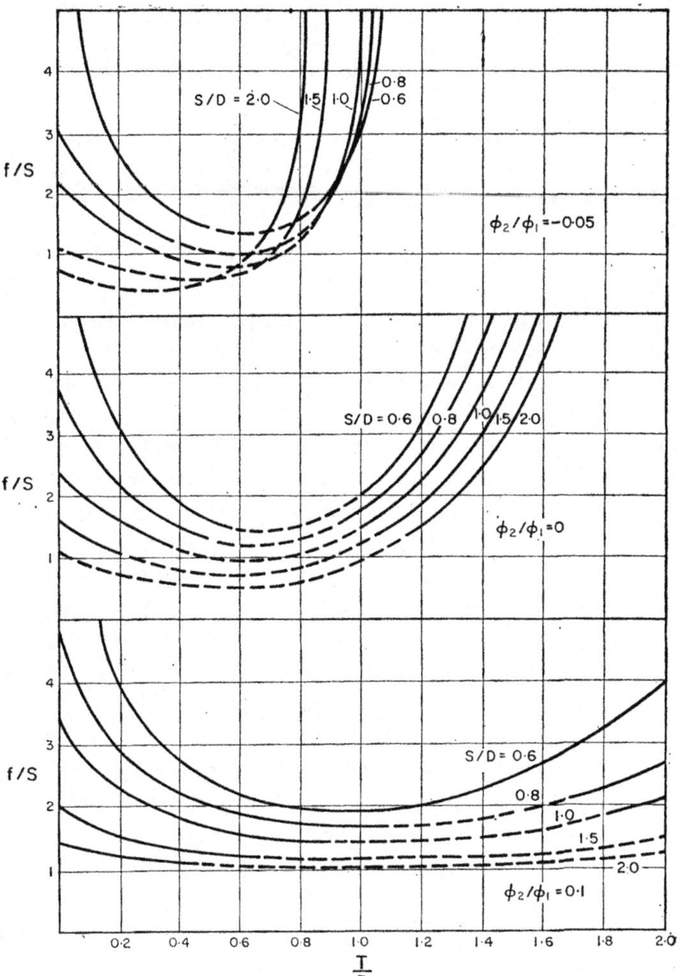

f/S
S/D = 2·0
1·5
1·0
0·8
0·6
φ2/φ1 = -0·05
S/D = 0·6
0·8
1·0
1·5
2·0
φ2/φ1 = 0
S/D = 0·6
0·8
1·0
1·5
2·0
φ2/φ1 = 0·1
0·2
0·4
0·6
0·8
1·0
1·2
1·4
1·6
1·8
2·0
T/D

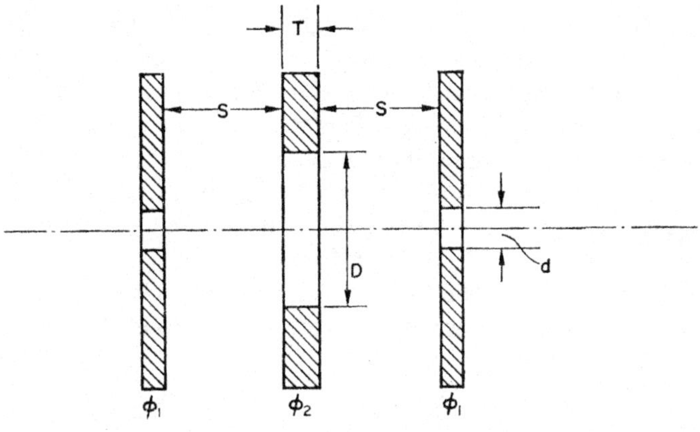

T
S
S
D
d
φ1
φ2
φ1

FIG. 17

information emerged from analyses by Regenstreif[51], and by Glaser and Schiske[17]. The latter was based on an assumed field distribution of the form

$$\phi = \phi_0 \left[1 + \frac{\alpha}{1+(z/L)^2} \right]$$

which applies when the central diaphragm is thin; this rigorous analysis was later extended by Archard[18] to thick diaphragms. The results, shown in Fig. 17, are taken from this paper, and relate to three different values of the central potential. The dotted portions of the curve indicate that the focal plane lies within the lens. Unlike the case of a magnetic lens, it is of course not possible to place the object near this focal point. However, the lens can be operated as a projector in this range. The curves neglect any dependence on the diameter of the aperture in the outside diaphragms. Field plots show that this dependence is quite small as long as the apertures are small compared with the diaphragm separations. The results presented in this figure have been shown to agree well with available experimental results[18].

Since it is the square of the magnetic field strength which determines the power of a lens [see eq. (2.51)], it is clear that a design in which the magnetic field is confined to the smallest possible region will, for a given lens strength, require the smallest magnetomotive force. For this reason almost all magnetic lenses used for imaging purposes have the general form shown in Fig. 18a. However, for velocity spectrometers, and for some image converters, one is also sometimes concerned with the focusing properties of unshielded coils of the form shown in Fig. 18b.

Our most detailed knowledge of the properties of such lenses undoubtedly comes from the work of Liebmann[52] based on the use of the resistive network for the determination of the field distribution, followed by numerical integration of the ray equation; we will here present some of his results. Since the primary data with which we are concerned are the focal length and position of the principal planes as a function of the lens strength, our first consideration must be to arrive at some suitable definition of "strength". Now, in section 3.1.4 it turned out to be useful to define the strength in terms of a quantity $(a/\rho)^2$ which related the extent of the field a to the radius of an electron ρ injected into a uniform magnetic field corresponding to the peak value attained in the lens. Now, for a lens such as that in Fig. 18a, both the peak of the distribution and its width are quantities which are known only after the field problem has been solved. It is therefore rather more convenient to define the magnetic field as

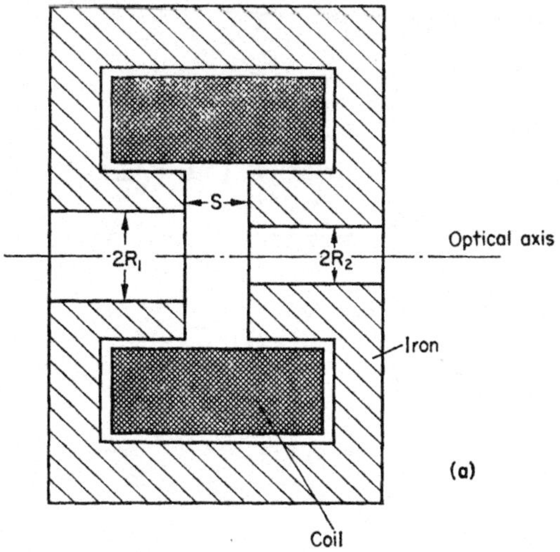

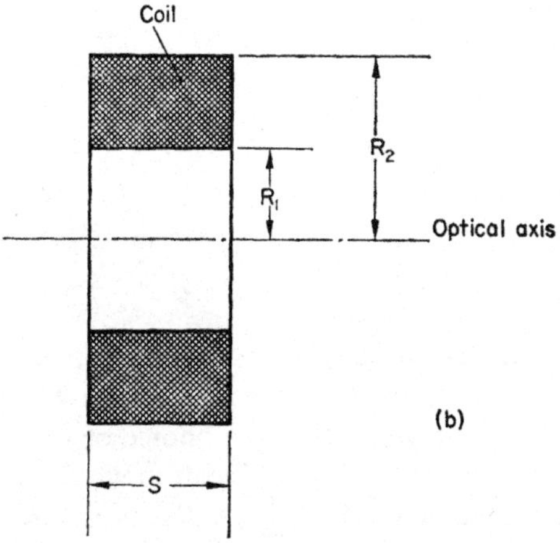

FIG. 18

that which would exist in the gap if we were to make $R_1 \ll S$ and $R_2 \ll S$. The extent of the field is defined in terms of one of the actual radii R_1. Now the magnetic field strength under these conditions, neglecting the reluctance of the iron, is simply proportional to $(NI)^2/S$, where NI is the total number of ampere turns in the coil. The parameter which would therefore correspond to $(a/\rho)^2$ would take the form

$$\frac{(NI)^2}{V}(R/S)^2.$$

In practice it has proved convenient to use just $(NI)^2/V$ as an even

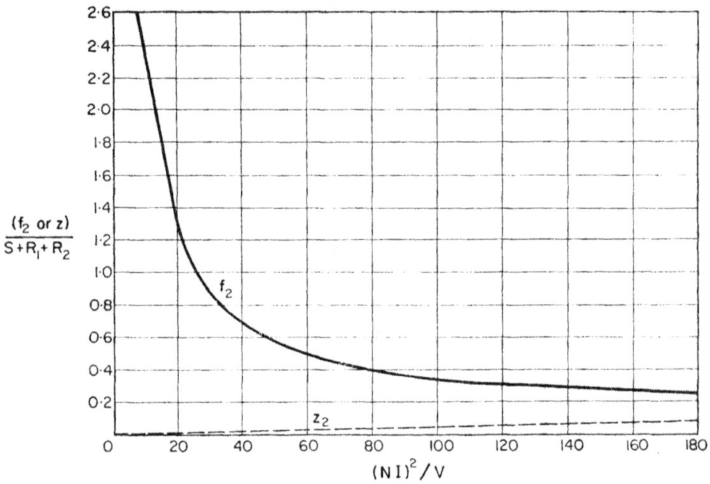

FIG. 19

simpler parameter. This is in effect a measure of the total magnetomotive force, "normalized" with respect to the beam voltage, a quantity which, for any *given* geometry, is a measure of lens strength of the same kind as $(a/\rho)^2$.

The results for focal length and principal plane position can, of course, be normalized with respect to any convenient distance variable. Liebmann discovered that the choice of $(S+R_1+R_2)$ for this variable led to a remarkable simplification which is shown in Fig. 19. It turns out that as long as $S/(R_1+R_2)$ is in the range of 0.5–2.0 a single curve can adequately represent all cases. Even for extreme, small or large, values of this ratio, the change in the location of the curve is not very large. Thus, for example, when the strength of the lens is 40 $[(NI)^2V=40]$, even for the extreme case of $S/D \ll 1$ or

$D/S \ll 1$, the focal length is only about 15 per cent longer than indicated by the curve in Fig. 19, which is valid for moderate values of the parameter. The percentage error increases for higher lens strength, and amounts to about 30 per cent maximum for a lens strength of 100. However, most practical lenses lie within a few per cent of the curve. Even the unshielded coil is moderately well represented by the curve if R_1 and R_2 are defined as in Fig. 18b. The focal lengths are somewhat longer, corresponding approximately to the case of the iron-shielded coil with very small or very large S/D. At the largest values of the lens strength shown in this figure the focal point is just beginning to enter the lens field. Thus, just as in section 3.1.4, we should under these conditions distinguish between the effective focal length when the lens is used with a virtual object (e.g. in the intermediate lens in an electron microscope) or if a real object is placed near the focus of the lens. The curve shown is for the latter case, i.e. for the case when the lens is used as an objective. However, even at the maximum strength shown the virtual object focal length is only about 15 per cent longer. At a lens strength of 100 the difference is quite negligible.

The principal plane is seen to move almost linearly from its central position at zero field strength. However, the shift is rather small, amounting to about 30 per cent of the focal length for the strongest length shown.

Figure 19 is indeed a remarkable example of compression of a very large body of significant data. It is possible to proceed a long way with the design of complicated apparatus involving magnetic optics, with Fig. 19 as the sole guide.

4. ABERRATIONS

The paraxial properties of lenses already discussed suffice for the design of electron optical devices for which the precision of imaging is not a critical requirement. However, more usually the lens aberrations must be considered in the design. It must be appreciated that by comparison with light optical lenses, the quality of electron lenses is very poor indeed; thus, for example, whilst in a light microscope the objective aperture will be of the order of unity, in the case of an electron microscope the spherical aberration is so marked that optimum resolution is obtained with an aperture of only around

10^{-3}. The electron microscope is, of course, the most critical of all applications, but it is found that even in the design of a comparatively primitive apparatus, it is sometimes important to take aberrations into account. In this section we will therefore give a concise account of the basic theory of aberrations, how they are calculated, and note some of the most important numerical data which are available.

We found in section 2.1, quite generally, that a force which varies linearly with radius from a symmetry axis will act as a perfect lens. In general, the variation of the force with radius must of course be represented by an infinite series such as eq. (2.1), which for circular symmetry is confined to odd powers of r. In developing the paraxial theory we expanded the potential and the magnetic field using three terms in the former [eq. (2.29)] and two in the latter [eq. (2.30)] since these alone determine the linear force field. Clearly, if we wish to calculate the deviations from the paraxial theory, we will have to take more terms into account. It should, however, be emphasized right away that this is by no means the only way to approach the matter. If, for example, we are concerned with a lens having given electrode shapes, we can compute the field at all points to which the electrons have access, then directly apply the general equation (2.26), in which no paraxial assumptions have been made and discover where a particular ray hits the image plane. In addition to the aberrations which arise from the inevitable departures of a law of force which is linear with distance from the axis, there will be aberrations arising from the fact that in a practical electron beam the electrons have a velocity distribution. The resulting "chromatic" aberrations will be discussed in section 4.4.

4.1. The aberration terms

The nature of the problem with which we are now concerned is explained in Fig. 20. A point in the object plane at a radius r_0 has a *paraxial* image point in the image plane at a radius r_i ($= Mr_0$). This point will in fact be the true image point if r_0 and r_0' are sufficiently small. However, if they are not, the true image point will be displaced by Δr_i. Now the paraxial image point is uniquely defined once the object point is given. However, the true image point is not—it depends also on the angle of the ray leaving the object point. We must therefore recognize that Δr_i is a function not only of r_0 but also of

$(r')_0$. Rather than developing a theory in terms of the dimensionally different quantities r_0 and $(r')_0$ it is more convenient to specify a ray in terms of r_0 and the radius at which it intercepts a second plane located at some distance from the object plane. The interception position on two planes is sufficiently unique to define the particular ray. It is clear that this additional plane, which is generally known as the aperture plane, can be placed at *any* convenient point, not excluding the region of the lens itself, or some position preceding the object

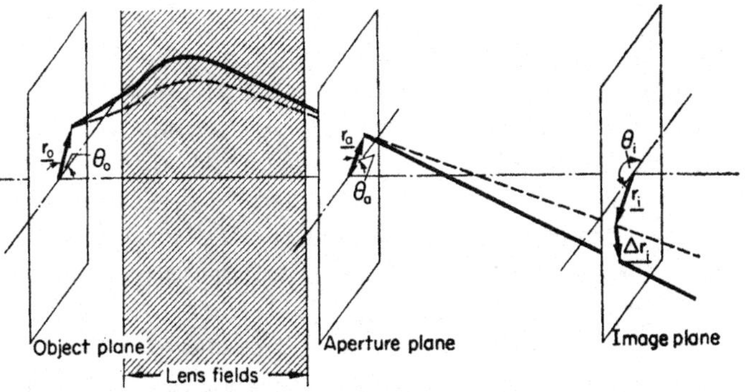

FIG. 20. Notation for discussion of aberrations. The broken line shows a ray in the paraxial approximation. The solid line is the actual trajectory, passing through the *same* points in both object and aperture planes, but deviates from the paraxial image point by the aberration vector $\Delta \boldsymbol{r}_i$.

plane or following the image plane. Largely for historical reasons it is usual to locate it at the point where the paraxial ray, with zero initial slope crosses the axis.†

In the case of magnetic lenses, the rays do not stay in a meridian plane [eq. (2.34)], the image being rotated with respect to the object. Thus the object and image positions should be regarded as vectors having both a magnitude and a polar angle:

$$\boldsymbol{r}_0 = r_0\, e^{j\theta_0} \tag{4.1}$$

† If this point should lie within the lens field, it is more usual to locate the aperture plane at the point where the ray projected back from the image point intersects the axis. This point will usually also lie within the lens field.

and similarly for $\boldsymbol{r}_a$ and $\boldsymbol{r}_i$. Moreover, the aberration distance itself, Δr_i, must also be regarded as a vector quantity

$$\Delta \boldsymbol{r}_i = \Delta r_i \, . \, e^{j\Delta\theta_i}. \tag{4.2}$$

In one sense, once we have obtained the aberration vector $\Delta \boldsymbol{r}_i$ as a function of r_0 and r_a our task is completed. However, for many purposes this general result will not be in a convenient form, unless the aberrations are too small to be of any great concern. If they are large, we would like to know whether we can in some way avoid them, or at least redesign the system to reduce their effect. The aberrations with which we have to contend might be due to the circumstance that, whilst the object to be imaged is quite small, some of the imaging rays are too far away from the axis in the lens. In this case we can obviously improve matters by inserting a limiting aperture (*not* necessarily in the "aperture" plane) to cut out those "far-out" rays which do most of the damage. On the other hand the dominant source of trouble may be that the object is large even though only those imaging rays in the lens field which are relatively close to the axis can reach the image. In this case further stopping down may not help. It is therefore of paramount importance to separate the aberrations into classes corresponding to some recognizable, and hopefully curable, defects of the optical system. It is this separation to which we will now turn our attention.

In discussing the relationship between $\Delta \boldsymbol{r}_i$ and $\boldsymbol{r}_0$ and $\boldsymbol{r}_a$, we can once again exploit the fact that we are concerned with a system having rotational symmetry. Now when we define our object point we say that its radius vector makes an angle θ_0 with respect to a particular, but *quite arbitrary* radius vector and similarly the radius vector to the intersection of the ray with the aperture plane makes an angle θ_a with the same reference vector, and similarly again for θ_i, at the image point. If now we rotate the reference vector through some angle ϕ, then θ_0, θ_a and θ_i must all change by ϕ.

The expression for r_i must take the form

$$\boldsymbol{r}_i = M\boldsymbol{r}_0 + \boldsymbol{f}(r_0, r_a), \tag{4.3}$$

where

$$\Delta \boldsymbol{r}_i = \boldsymbol{f}(r_0, r_a).$$

Now suppose we were to assert (falsely) that $f(r_0, r_a)$ contains a term, such as (const.)$r_0 . r_i$. We would then deduce that rotation of the reference vector through ϕ would rotate both r_0 and r_i through ϕ and hence Δr_i through 2ϕ—suggesting that a mere change of co-ordinates had changed the physical rotation of the ray in the lens. Similarly, if we were to try a term such as $r_0 r_a^*$ (where r_a^* is the complex conjugate of r_a) we would find that the rotation of Δr_i was zero instead of ϕ. Thus we see that no even products of r_0 and r_a can occur, and that the next higher terms in $f(r_0, r_a)$ must be *triple products*. Here again, the same argument allows us to eliminate a number of possibilities. For example, if we had a term r_0^3, Δr_i would be rotated through 3ϕ when the object is rotated through ϕ. However, a product such as $r_0 r_0 r_0^*$ will lead the required image rotation of ϕ. If we assemble all the triple products which satisfy this requirement, we obtain the expression

$$\Delta r_i = F r_0^2 r_0^* + E r_0 r_0^* r_a + D r_0^2 r_a^* + C r_0 r_a r_a^* + B r_0^* r_a^2 + A r_a^* r_a^2. \quad (4.4)$$

This is as far as we can go in complete generality. If, however, we are prepared to accept a constraint on the (otherwise quite arbitrary) position of the aperture plane, we can obtain a further simplification. The necessary constraint is to ensure that the aperture plane lies beyond the region of significant lens fields, so that the trajectories between it and the image plane are straight lines. For such a reason it can be shown[(68)] that the trajectories form a series of normal rays to a wave front. Purely geometrical considerations then lead to the result that the imaginary part of $d(\Delta r_i)/dr_a$ must vanish. Applying this condition to eq. (4.4), and noting that since r_0 and r_a are arbitrary, it must apply equally to each term separately, we find that A must be real, $C = 2B^x$ and E must be real, so that we may rewrite equation in the form

$$\Delta r = |A| r_a^* r_a^2 + B r_a^2 r_0^* + 2_0 B^* r_0 r_a r_a^* + D r_0^2 r_a^* + |E| r_0 r_0^* r_a + F r_0^2 r_0^*. \quad (4.5)$$

Each of the terms in this equation represents a distinct and recognizable type of aberration. In view of the complete generality in its derivation it is clear that these must correspond to the various well-known optical aberrations, and in the case of the electrostatic lens (for which all the coefficients are real) the correspondence is complete. In eq. (4.5) the first terms (coefficient A) is spherical aberration, the second and third (B and B^*) is coma, the fourth astigmatism, the fifth curvature of field and the last (F) distortion. In the case of a magnetic lens the circumstance that some of the coefficients are now complex leads to an additional "twist" superimposed on the aberrations. In a later section we will give a brief description of the physical nature of these aberrations, and how they reveal themselves in practice, but before embarking on this, we will consider how the coefficients in eq. (4.5) are to be calculated.

4.2. Calculation of the aberration coefficients

The aberration coefficients may be derived from the general ray equation (2.3.6) by a direct extension of the technique which led to the paraxial equation; by including higher order terms in the expansion of the fields, we can obtain an equation which is accurate to the same order as eq. (4.5), i.e. including third order terms but neglecting fifth order. The additional terms can be regarded as a perturbation of the original paraxial equation, and analytical solutions for the coefficients can be derived—although they tend to be rather involved. This general approach, which was pioneered by Scherzer[(54)], has formed the core of aberration theory ever since. The value of this technique lies in the possibility which it offers of obtaining general results and, possibly, of providing the basis of a synthesis technique[(55)] which can lead to the design of improved lenses. One very important result, obtained by Scherzer[(54)] is that by suitable combination of fields all lens errors can be eliminated, with the exception of spherical aberration.† This result is restricted to lenses having circular symmetry and free from space charge; both these restrictions provide a possible escape from the conclusion, and will be briefly discussed later.

The analytical expression can, of course, also be used for the numerical evaluation of aberrations from an assumed or measured field distribution. Here, however, the method must compete with the direct attack based on the integration of the ray equations. In the following two sections both approaches will be outlined.

4.2.1. PERTURBATION THEORY OF ABERRATIONS

In the following we will confine ourselves to the electrostatic case. The presence of magnetic fields makes the algebra worse but does not involve any essential complication.

† . . . and also of chromatic aberration which will be discussed in section 4.4.

We start with the expansion of the potential given in eq. (2.29). Unlike the paraxial case we will on this occasion retain the term in r^4. Thus we will take

$$\begin{aligned} \Phi &= \phi - \frac{r^2}{4}\phi'' + \frac{r^4}{4}\phi'''', \\ \frac{\partial \Phi}{\partial r} &= -\frac{r}{2}\phi'' + \frac{r^3}{16}\phi'''', \\ \frac{\partial \Phi'}{\partial z} &= \phi' - \frac{r^2}{4}\phi'''. \end{aligned} \tag{4.6}$$

When these are inserted into eq. (2.15) the result may be expressed in the form

$$r'' + r'\frac{\phi'}{2\phi} + r\frac{\phi''}{4\phi} = \frac{r^3}{16}\left[\frac{\phi''''}{2\phi} - \left(\frac{\phi''}{\phi}\right)^2\right] + \frac{r^2 r'}{8}\left[\frac{\phi'''}{\phi} - \frac{\phi'\phi''}{\phi^2}\right] - \frac{rr'^2}{4}\frac{\phi''}{\phi} - \frac{r'^3}{2}\frac{\phi'}{\phi}. \tag{4.7}$$

Here we have separated the terms which make up the paraxial equation on the left-hand side. This suggests that one might put

$$r = r_p + \Delta r, \tag{4.8}$$

where r_p is the solution to the paraxial equation, and one finds,

$$\Delta r'' + \Delta r'\frac{\phi'}{2\phi} + \Delta r\frac{\phi''}{4\phi} = \frac{r_p^3}{16}\left[\frac{\phi''''}{2\phi} - \left(\frac{\phi''}{\phi}\right)^2\right] + \frac{r_p^2 r_p'}{8}\left[\frac{\phi'''}{\phi} - \frac{\phi'\phi''}{\phi^2}\right] - \frac{r_p r_p'^2}{4}\frac{\phi''}{\phi} - \frac{r_p'^3}{2}\frac{\phi'}{\phi}. \tag{4.9}$$

It is important to appreciate that since r' satisfies the paraxial ray equation, the right-hand side of eq. (4.9) can be written in a variety of equivalent forms by making use of this fact. This equation which is, of course, linear in Δr can be solved for Δr. The solution takes the form of a definite integral over the complete lens fields of a rather complicated function of the lens fields, including derivatives up to the fourth, and the solution to the paraxial equation r_p with its derivatives. Moreover, this integral can be manipulated to appear in the same form as eq. (4.5) so that the individual coefficients can be recognized and separately expressed. The explicit expressions for these coefficients, which are rather lengthy, are listed in a number of texts[56], and will not be reproduced here. It is felt that their main value lies in the general results which Scherzer was able to obtain from them rather than for the actual computation of errors, which it is easier and quicker to obtain either by the use of eq. (2.86) or by the direct integration of the ray equation (4.7). However, it should be noted that it is possible to use the analytical aberration expressions as the basis of a variational calculation for optimizing specific lens parameters. Some work in this direction has been reported, for example by Trettner[55], and appears to be a powerful tool for certain applications.

4.2.2. DIRECT NUMERICAL EVALUATION OF ABERRATIONS

If we integrate eq. (4.7) on a computer, we can obtain the total aberration Δr_i directly. In some cases this information may suffice. For example, if we are trying to devise an electron optical system to ensure that an electron beam is able to pass through a small aperture with very little interception, our interest will be satisfied if we find that Δr_i is less than the radius of the aperture for most typical trajectories. However, if we want a more detailed picture of the aberrations, it is necessary to find some means of determining the actual aberration constant appearing in eq. (4.5). One can readily outline the steps which are required to do this.

The first is to trace a paraxial ray to determine the image point for a particular object position. To do this one can use the paraxial equation (2.36) or use the complete equation (4.7) (which will in any event have to be programmed), and simply use very small values of r_0 and r_a. The next step is to trace a ray for an object on the axis, passing through the aperture plane at a radius r_a. Under these conditions only the first term in eq. (4.5) contributes to the aberration, and we obtain A directly in terms of the computed error. We can then trace a ray starting at a radius r_0 and crossing the axis in the aperture plane. The complex error Δr_i will then give us the real and imaginary parts of F.†

† In practice one would programme the problem for initial conditions r_0 and r_0'. The trajectory for which $r_a = 0$ would then be found by successive approximations.

The determination of the other coefficients can be effected by tracing three further rays. Let the object position be real, so that

$$\boldsymbol{r}_0 = r_0$$

and let us write

$$\boldsymbol{r}_a = r_a \exp(j\theta_a)$$

and

$$\boldsymbol{B} = B_r + j\,B_i, \text{ etc.}$$

Then, for a particular (and arbitrary) choice of the magnitudes of $|r_0|$ and $|r_a|$, let the error in the image plane be denoted by $\Delta_{\theta a}$. A convenient choice for the three values of θ_a is 0, π and $\pi/2$. From eq. (4.5) we then have

$$\begin{aligned}
\Delta_0 &= Ar_a^3+(B+2B^*)\,r_a^2 r_0+(D+E)\,r_0^2 r_a+Fr_0^3,\\
\Delta_\pi &= -Ar_a^3+(B+2B^*)\,r_a^2 r_0-(D+E)\,r_0^2 r_a+Fr_0^3,\\
\Delta_{\pi/2} &= +jAr_a^3-(B-2B^*)\,r_a^2 r_0+j(-D+E)\,r^2 r_a+Fr_0^3.
\end{aligned} \tag{4.10}$$

We can readily solve these for the B, D and E in terms of the Δ and the coefficients A and F already determined:

$$\begin{aligned}
B_r &= \frac{1}{6r_0 r_a^2}\ \mathrm{Re}\,[\Delta_0-\Delta_\pi-2Fr_0^3],\\
B_i &= -\frac{1}{2r_0 r_a^2}\ \mathrm{Im}\,(\Delta_0-\Delta_\pi-2Fr_0^3],\\
D_i &= \frac{1}{2r_0^2 r_a}\ \mathrm{Im}\,[\Delta_0-\Delta_\pi],
\end{aligned}$$

$$D_r = \frac{1}{4r_0^2 r_a}\,\{\mathrm{Im}\,[3\Delta_0-\Delta_\pi-2\Delta_{\pi/2}-4Fr_0^3]+\mathrm{Re}(\Delta_0-\Delta_\pi)\}, \tag{4.11}$$

$$E = \frac{1}{4r_0^2 r_a}\,\{\mathrm{Re}\,(\Delta_0-\Delta_\pi)-\mathrm{Im}\,[3\Delta_0-\Delta_\pi-2\Delta_{\pi/2}-4Fr_0^3]-4Ar_a^3\}.$$

4.3. Physical nature of the aberrations

In general a lens will have all the defects represented by the various terms in eq. (4.5). It is nevertheless worth while to consider the various defects separately, for in many situations one or two of the terms will dominate the aberration, or at any rate be of dominant importance in a specific application. Thus, spherical aberration is of supreme importance in electron microscopy, and distortion a comparatively minor consideration. On the other hand in a cathode ray oscilloscope designed for precision measurement, the avoidance of distortion is of paramount importance.

One could discuss the nature of each aberration from a number of different points of view. However, one approach which has its roots in light optics is to consider the form of the "aberration figure". This figure is produced by noting the interception on the image plane for a series of rays coming from a particular object point, when the rays are chosen so as to hit the aperture plane at various points on the circle of radius r_a. The aberration figure is therefore the image which would be observed for a point object, when the aperture consists of a thin annulus of radius r_a. Indeed, it is in effect the analysis of the aberration figure which we used in the last section, in the method which we suggested for computing the aberration coefficients. The emphasis on the aberration figure might seem slightly abstract at first sight. Its use was, however, first suggested by the fact that it is a diffuse form of this figure which can often be observed in practice, even though a circular rather than annular aperture would normally be involved.

In the following we will briefly discuss the aberration figures for the various aberrations, but also indicate the form of the image of a complete object which would be observed in the presence of the particular aberration considered.

4.3.1. SPHERICAL ABERRATION

From eq. (4.5) we see that

$$\begin{aligned} \Delta r_i &= A\, r_a^* r_a^2 = A\, r_a^3 \exp(j\theta_a), \\ |\Delta r_i| &= A\, r_a^3, \\ \theta_i &= \theta_a. \end{aligned} \tag{4.12}$$

Figure 21 shows the origin of the aberration figure. If we consider a series of rays from one object point such that they trace out the circle of radius r_a in the aperture plane, the trace on the image plane is a circle of radius Ar_a^3, one revolution, in a particular sense, for θ_a corresponding to one revolution, in the same sense, for θ_i.

Spherical aberration has probably received more attention than any of the rest, because it is the only one that afflicts an object point, even when it is on the axis. It therefore represents an obstacle to increasing resolution, even for very restricted object sizes. Each annular portion of the total imaging bundle forms a focus at a different point along the axis, and leads to the circular aberration figure shown. The combination of all the annuli leads to the appearance of the shaded circular disc centred on the Gaussian image point. As a result, if we image an object such as shown in Fig. 21b, the resulting image is as shown in Fig. 21c, the degree of blurring being the same

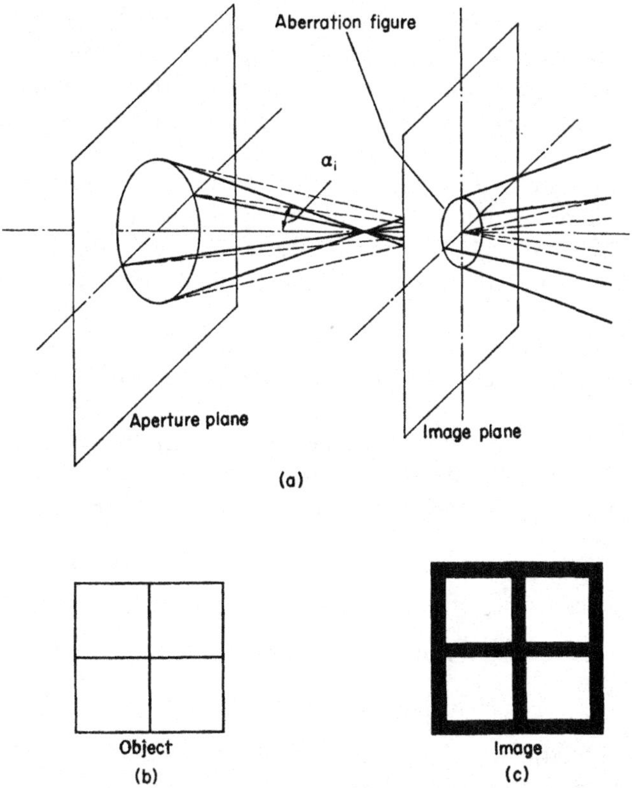

FIG. 21. Spherical aberration. (a) Origin of aberration figure. (b) Test object. (c) Image of object.

for all object points. As shown by Scherzer(54), the sign of the spherical aberration is always as indicated in the figure.

It is important to appreciate that the numerical value of A is not an exclusive property of the lens—it depends also on the choice of the position of the aperture plane even though this is involved in our present discussion only as a part of the coordinate system for defining a particular ray. In discussing the quality of a particular lens it is, of course, desirable to describe the aberration by means of a parameter which depends on the lens alone. One can avoid any reference to the "aperture plane" by working in terms of the angle of arrival at the image plane α_i, and write

$$\Delta r_i = C_s \alpha_i^3. \tag{4.13}$$

Here C_s has the dimensions of length and is sometimes expressed in terms of the focal length of the lens $C_s' = C_s/f$. However, most numerical results which have been published are in terms of C_s so that it is this parameter rather than the non-dimensional C_s' which we will use.

We have so far discussed what goes on at the image plane—i.e. the plane where, in a more perfect world, we would find the undistorted image of the object. There is, of course, no need to observe the distorted image at this particular plane. Indeed, in practice its very location may be far from obvious. One is rather concerned with finding the plane where the aberrations do least damage. In general this will not coincide with the Gaussian image plane. In the case of spherical aberration, the aberration disc has a minimum diameter which will be somewhere between the Gaussian plane and the plane where the extreme rays are brought to a focus. The location of the "disc of least confusion" is discussed in many texts on optics.

4.3.2. COMA

Two of the terms in eq. (4.5) contribute to coma:

$$\Delta r_i = B r_a^2 r_0^* + 2 B r_0 r_a r_a^*. \tag{4.14}$$

We will for the moment treat the case when B is real, which applies to electrostatic lenses. For this case

$$\Delta r_i = B r_0 r_a^2 [\exp(2j\theta_a) + 2]. \tag{4.15}$$

Thus we see that the aberration figure is again a circle, but this time displaced by a definite amount $(2Br_0 r_a^2)$ from the Gaussian image point. Moreover, for one revolution of the trace on the circle radius r_a in the aperture plane, the image point carries out two revolutions over the aberration figure. Figure 22 illustrates this behaviour; the eight rays equally spaced on the aperture circle are seen to impinge on the aberration figure at four equally spaced points. The various aberration figures corresponding to different values of r_a are always tangential to the two straight lines enclosing an angle of $\pi/6$, which are shown on the figure. These lines are symmetrical with respect to the meridian line for this case. The sole effect of a complex

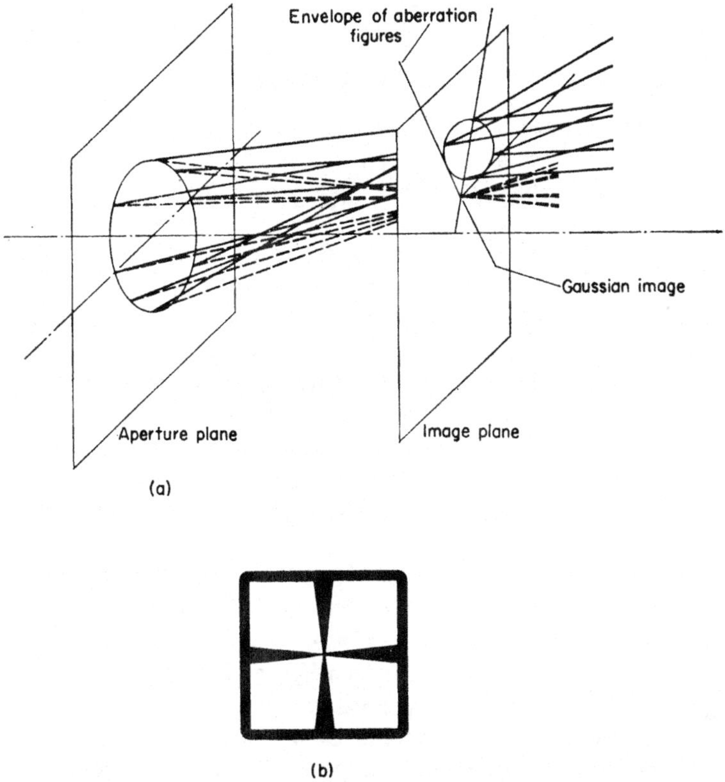

FIG. 22. Coma (isotropic). (a) Origin of the aberration figure. (b) Appearance of test pattern, centred on axis.

B, such as encountered in magnetic lenses, is to tilt the central line of the aberration figures with respect to the meridian.

The appearance of the test object shown in Fig. 21b when imaged by a system with isotropic coma as the sole aberration would take the form sketched in Fig. 22b.

Coma is likely to be troublesome when large imaging apertures are used, even with a relatively small object. This kind of situation is sometimes encountered when attempting to achieve relatively high current densities with low voltage beams, for in this case, as will be discussed in section 6.1, the use of large apertures is demanded by the need to avoid the deleterious effects of the thermal velocity distribution.

4.3.3. ASTIGMATISM

The term in eq. (4.5) which gives rise to astigmatism is

$$\Delta r_i = D r_0^2 r_a^* = D\, r_0^2 r_a \,.\, \exp(-j\theta_a). \qquad (4.16)$$

Once again we will confine ourselves largely to the case when the coefficient is real. The aberration figure is again a circle centred on the Gaussian image point. However, here, if the rays trace out the circle radius r_a on the aperture plane in one sense, the aberration figure is traced out in the *opposite* sense. This has a very significant effect on the behaviour of the bundle of rays, which is shown in Fig. 23a. We have shown two separate bundles, having different radii at the aperture plane. We see that the rays in one plane come to a focus one side of the image plane, and those in the orthogonal plane are focused on the other side. This partial focusing takes place for all rays in the plane, *irrespective of their radius at the aperture plane.* The position of the focus for the tangential or the meridianal rays depends only on the value of r_0—i.e. the position of the object point. The effect of this can be seen by comparing Fig. 23b and c. Figure 23b shows the general appearance of the test object in the Gaussian image plane. Figure 23c shows the image which would be observed in that plane where the rays emanating from a point on the object at a radial distance of r_{01} from the object have a radial focus. Corresponding pictures can be drawn for a plane where some of the tangential parts of the figure would be in focus.

Astigmatism depends on r_0^2 and is therefore normally a problem for extended objects—as long as the system is strictly circularly symmetric. However, if there is any departure from symmetry, severe astigmatism may be encountered even for objects on the nominal axis. Physically, we can regard astigmatism as arising from the fact that even with circular symmetry, once one is off the axis all directions are no longer equivalent; loss of circular symmetry leads to this same situation even on the axis. The understanding of this fact and its cure were of great importance in the perfection of the electron microscope.

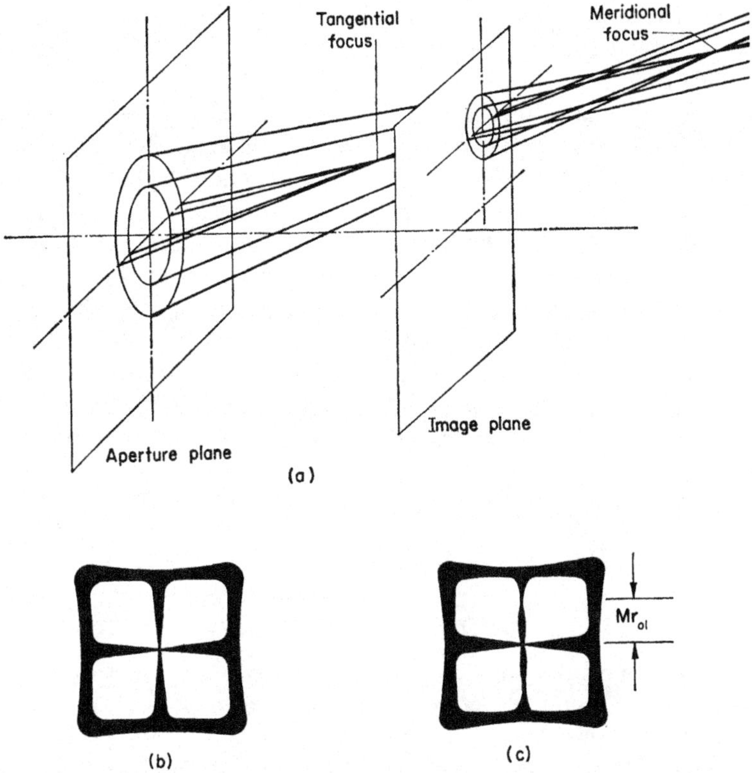

FIG. 23. Astigmatism (isotropic). (a) Aberration figures for two values of r_a. The foci for rays having different values of r_a coincide. (b) Image of test object at the Gaussian image plane. (c) Image of test object located at point where an object point at a radius of r_{01} develops its meridianal focus.

In the case of magnetic lenses, D will be complex. This leads to the appearance of partial (line) foci not for the radial and tangential directions, but for two other directions, which may have any angle with the meridian plane, but which are still orthogonal to each other.

4.3.4. CURVATURE OF FIELD

$$\Delta \boldsymbol{r}_i = \boldsymbol{E} \boldsymbol{r}_0 \boldsymbol{r}_0^* \boldsymbol{r}_a = \boldsymbol{E} r_0^2 \boldsymbol{r}_a \exp(j\theta_a). \qquad (4.17)$$

This term is therefore very similar to astigmatism, differing from it only in the sign of θ_a. Thus a rotation in one sense of the point on the circle of radius r_a in the aperture plane leads to a rotation in the same sense on the aberration figure.

This difference, however, leads to a totally different behaviour. It turns out that in this case, for any specified value of r_0 there is a point in space where all rays, independent of their value of r_a, will come to a focus. It is apparent from eq. (4.17) that the locus of these points must be a surface which departs from the Gaussian image plane by an amount proportional to r_0. It is therefore a parabolic surface, which in electron optics is invariably concave towards the lens. We can therefore effectively eliminate the loss of sharpness arising from the aberration, simply by giving the observation plane† a suitable curvature. Thus whilst the appearance of the image of the test object at the image plane would be precisely as shown in Fig. 23b, on the appropriate parabolic surface the image would be sharp.

4.3.5. DISTORTION

$$\Delta \boldsymbol{r}_i = \boldsymbol{F}.\boldsymbol{r}_0^*.\boldsymbol{r}_0^x = \boldsymbol{F} \boldsymbol{r}_0^3. \qquad (4.18)$$

Distortion is the simplest of all the aberrations to picture. It depends only on the object point and in no way on where a ray intersects the aperture plane. Thus all rays from an object will recombine to form a sharp image, which will, however, be displaced from the Gaussian image point. This displacement can have either sign, and for a magnetic lens, for which F will in general be complex, there will also be a twist.

Figure 24 shows the form of the image of the test object for the two signs of F (assumed real). Distortion is particularly important in any electron optical device where a wide field of view is essential. It is therefore of importance in cathode ray tubes for scientific and picture applications and also in devices such as storage tubes and

† Or the object plane—a remedy which is frequently applied to devices such as image convertors.

image converters. It is one of the defects which can be eliminated by sufficiently careful design.

4.4. Chromatic aberration

A prerequisite to the perfection of the imaging under paraxial conditions is an electron beam which is perfectly mono-energetic. In practice there are two factors, very different in origin, but similar in effect, which introduce an effective spread in energy. The first is

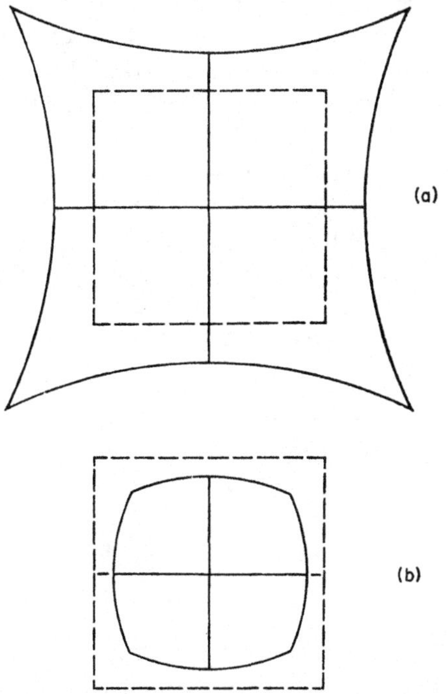

FIG. 24. Distortion. (a) "Pin cushion." (b) "Barrel."

the velocity spread of electrons which is inevitable whether the cathode is thermionic, photoelectric or secondary emission. This velocity spread may be further enhanced if, as in an electron microscope, the electrons can lose energy in inelastic collisions. The second factor is due to the inevitable fluctuations in the supply voltages to the electron gun and to the lens electrodes or coils. The effect, in both cases, can be thought of as resulting in a variation of the refractive index of the medium, the variation being from one electron to the next for velocity spread, whilst supply fluctuations give a time variation.

It needs no more than a recognition of the fact that it is harder to deflect fast electrons than slow electrons to appreciate that the focal length of all lenses must increase with electron energy. In addition to this variation of effective focal length, one can also recognize two other aberrations: the lateral magnification will be a function of electron energy as will be the image rotation in magnetic lenses.

The theory of chromatic aberration is very much simpler than that for the geometric aberrations, because we are concerned with a very simple perturbation of the paraxial equation. The expressions for the aberrations once again involve integrals over the lens region of functions which involve the fields and the paraxial solutions. For computational purposes, the direct integration of the paraxial equation for two different initial energies is once again at least as convenient. However, the perturbation calculations do lead to a very simple and useful result for the case of thin lenses. This result is made the more useful by the fact that, as proved by Glaser[(58)], it represents an upper limit to the error. The expressions are concerned with shift in the focal length Δf.

For the magnetic lens

$$\frac{\Delta f}{f} = \frac{\Delta\phi}{\phi} - 2\frac{\Delta B}{B}. \tag{4.19}$$

For the electrostatic lens, under conditions of large magnification,

$$\frac{\Delta f}{f} = 2\frac{\Delta\phi}{\phi^{1/4}\,.\,\phi_m^{3/4}}. \tag{4.20}$$

In eq. (4.19) ΔB represents a fluctuation in the peak value of the magnetic field B. In eq. (4.20) ϕ is the potential on either side of the lens assumed to be the same (or the arithmetic mean if they differ), and ϕ_m is the minimum value in the lens field. Although the expressions represent upper limits it would appear that in practice it is not usually possible to realize values which are lower by more than a factor of around 0.7. This factor [by which the error is below that given in eq. (4.19)] is often referred to as the chromatic aberration coefficient and is usually denoted by C_c.

Unlike C_c it is possible to compensate for errors of chromatic *magnification* by a suitable combination of lenses. This problem is of some importance in electron microscopy and has been discussed by Weitzch[59].

Finally, it should be noted that the chromatic "error", which in most applications we would strive to reduce as far as possible, is the very thing which we seek to foster in another—in velocity spectrometers. The aim here is to separate different velocity classes so that a large chromatic aberration represents a figure of merit.

4.5. Reduction and compensation of aberrations

Much of the science of light optics has traditionally been concerned with the elimination of various aberrations. The spectacular success which has been achieved is based on the availability of several different media, having markedly different dispersive characteristics, and the freedom to shape the optical surfaces. In electron optics, we have, in one sense, a very direct control of the medium—we can change its "refractive index" at will—but on the other hand can do little to control the dispersion. The most serious limitation is that the shapes of the refractive surfaces—the equipotentials—are subject to the restrictive condition that they must satisfy Laplace's equation. The full import of these restraints was investigated in the classic paper by Scherzer[54] already mentioned. He was able to prove that no combination of electric and magnetic fields could reduce either the spherical or the chromatic errors, C_s and C_c, to zero—or to reverse their signs. Other errors including coma and astigmatism could in principle be corrected. This result represents the most important (perhaps the only important) fruit of the perturbation theory of aberrations. It is a result obtained on the assumption of steady, rotationally symmetric fields, subject to Laplace's equation.

In principle, as shown by Gabor[60], spherical aberration could be overcome by incorporating space charge regions in the electron optical system. This offers the designer the much greater flexibility of Poisson's equation instead of Laplace's equation. A number of ways of incorporating space charge have been suggested. A successful attempt to achieve correction in this way has been reported by Haufe[61], but the experiments were not concerned with a high precision objective such as used in electron microscopy, where the incentive for a reduction of spherical aberration is greatest. Indeed, it has been shown[62] that in this, the most critical of all the possible applications, space charge correction is not possible in principle, as the deleterious effect of electron interaction scattering outweigh the advantages of reduction in C_s. It was pointed out by Scherzer[63] (and at almost the same time by Hillier *et al.*[64]) that by abandoning rotational symmetry spherical aberration can, in principle, be corrected. This suggestion has been followed by a number of workers[65], and is currently still being actively investigated[66]. The prospects of this attack for reducing the aberration of a high precision objective are not very favourable. However, there is a range of other applications where a reduction in aberration would provide considerable design advantages. The chances of success in this field appear very much more favourable.

The Scherzer non-rotationally symmetric field technique still retains a straight optical axis. Gabor[67] has examined the possibilities which appear when this restriction is removed. He finds that chromatic error of aberration can be eliminated, or enhanced. The latter effect has been exploited in a new form of velocity spectrometer[68].

A further possible avenue for compensating spherical aberration and chromatic aberration is by the use of electron mirrors, which by dint of returning the electrons in the direction from which they came, manage to escape from Scherzer's judgement. Mirrors will be discussed in section 5. For the present one can summarize by saying that they are useful for special purposes but unlikely to cure aberrations in a high resolution device such as the electron microscope. They have, however, been successfully used to compensate chromatic aberration in image convertors.

Finally, we should mention what is perhaps the most obvious way of all to escape from Laplace's equation—by using electron permeable membranes or grids to define equipotential surfaces "artificially". Such membranes were indeed used in the first electrostatic lenses ever devised[69]. Once again the conclusion is that in practice they would do more damage than they could cure in a high resolution electron optical apparatus. However, a grid has recently been used with very great advantage in order to control the field distribution in a precision cathode ray tube[70]—a good example of the rebirth of a long dormant technological idea.

In electron microscopy the direction of progress has been largely found by concentrating on a reduction of spherical aberration rather than its compensation. This leads to the use of short focal length lenses, using the highest possible field strengths consistent with available materials. In the case of electrostatic lenses, the limiting factor is vacuum breakdown. Whilst a good deal has been learnt in recent years on how to increase the maximum safe field strengths under ideal conditions of surface preparation and ultra high vacuum, the safe limit in a working electron microscope where ideal conditions can hardly be maintained is unlikely to increase very much. Improved magnetic materials have helped in achieving a modest improvement

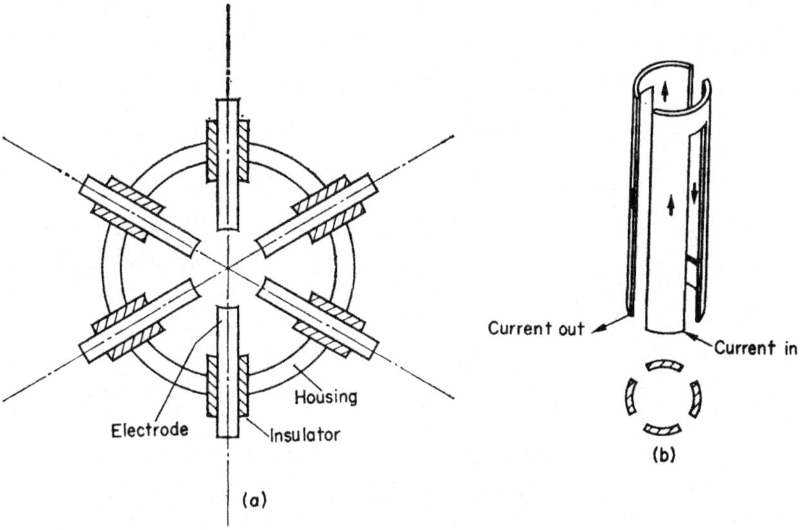

FIG. 25. (a) Electrostatic. Potentials to opposed electrodes are equal. Magnitude and sign for potential on each of the three opposing pairs separately adjustable. (b) Magnetic. Current flows in same sense in opposing legs. Mechanical rotation controls direction of correction.

in maximum field strengths. Cryogenic magnets offer at least a possibility of much stronger lenses, but there are great difficulties in practical implementation.

The astigmatism error can be largely compensated by the use of a special two-pole (i.e. cylindrical) lens. As noted in section 4.3.3, the main source of this aberration usually lies in an imperfection of the circular symmetry rather than in the "natural" astigmatism of the lens. The fact that the best machining techniques involve tolerance errors which still degrade the performance of the lens on this count was first discovered by Hillier[64], who obtained a great improvement by incorporating a system of iron screws which formed a weak cylindrical magnetic lens. Present-day practice for magnetic lenses is to use an electrostatic quadrupole field[71] or a magnetic quadrupole field[62], the former would normally be used for electrostatic lenses. In order to achieve complete compensation the magnitude and the direction of the quadrupole field must be adjustable. The latter can be achieved by providing a facility for mechanical rotation of the corrector, but is more conveniently obtained by the use of a more complicated series of electrodes which allow the rotation to be effected by controlling potentials or currents. Figure 25a shows an electrostatic corrector used in an electrostatic microscope. Fig. 25b shows the magnetic corrector described in ref. 72.

4.6. Numerical values of spherical and chromatic aberrations

The amount of numerical data on lens aberrations which has been published is not very large, particularly when compared with the volume of related algebra. The main body of results naturally relate to electron microscope lenses where the need to know is greatest. However, now that complete solutions can be readily obtained by purely digital means, aberrations of lenses used in less demanding applications can be undertaken. A recent example is the determination of the astigmatism in an infrared image converter.

The most important source of data for magnetic lenses is the analytical results obtained by Glaser for the bell-shaped distribution, and the resistance network results published by Liebmann[52]. As in the case of the focal properties Liebmann found that for almost the whole range of the geometric parameters which are of interest, the essential results could be condensed into a single curve for the spherical aberration constant and one for the chromatic constant. The limits of validity are much as discussed in section 3.2.3. It is worth adding that as far as the resolution of an electron microscope is concerned, the constant C_s enters into the expression by the one-fourth power. It is not therefore vital to know the value with very great precision. The results obtained by Liebmann, in the form presented by Haines, are shown in Fig. 26 for C_s and in Fig. 27 for C_c. It is clear from both curves that the aberrations will be minimized for the shortest possible focal length lens. These curves cannot, of course, show where the minimum focal length occurs—this depends on the materials properties already mentioned. For a maximum field strength of 20 kg, corresponding to the safe limit for currently available magnetic materials, the minimum value of $f/(S+R_1+R_2)$ is of

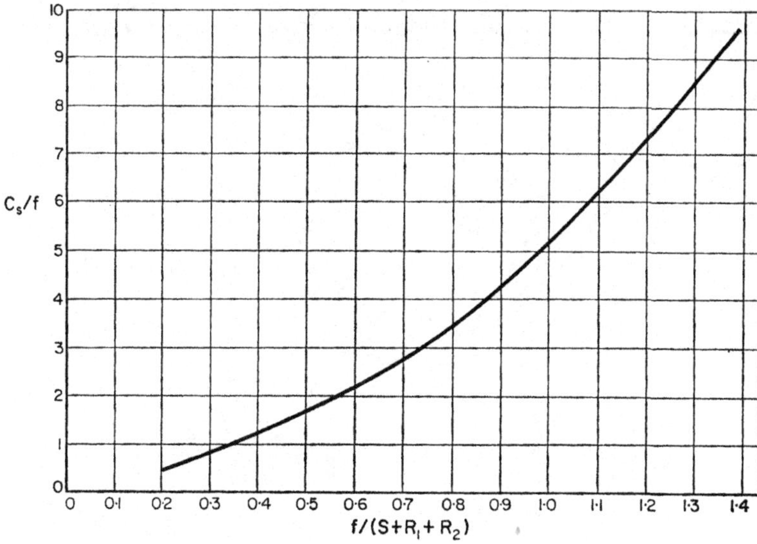

FIG. 26

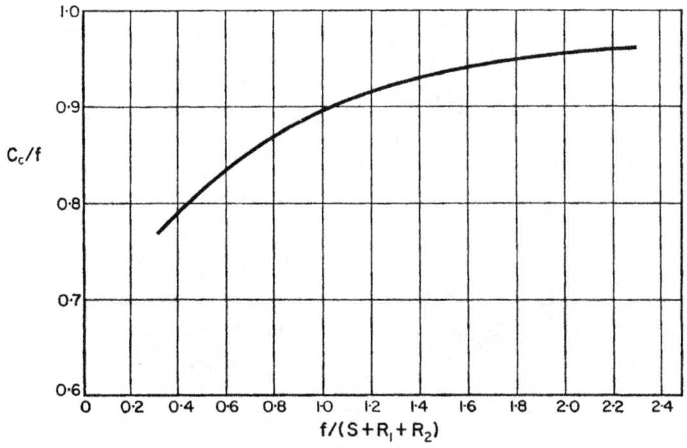

FIG. 27

the order of 0.3. The significance of the geometrical parameters was explained in Fig. 18.

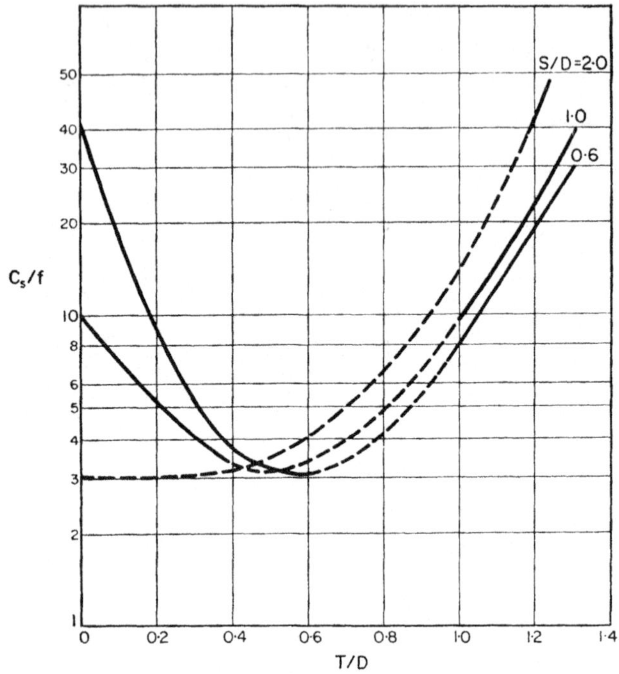

S/D=2·0
1·0
0·6
50
40
30
20
C_s/f
10
8
6
5
4
3
2
1
0
0·2
0·4
0·6
0·8
1·0
1·2
1·4
T/D

FIG. 28

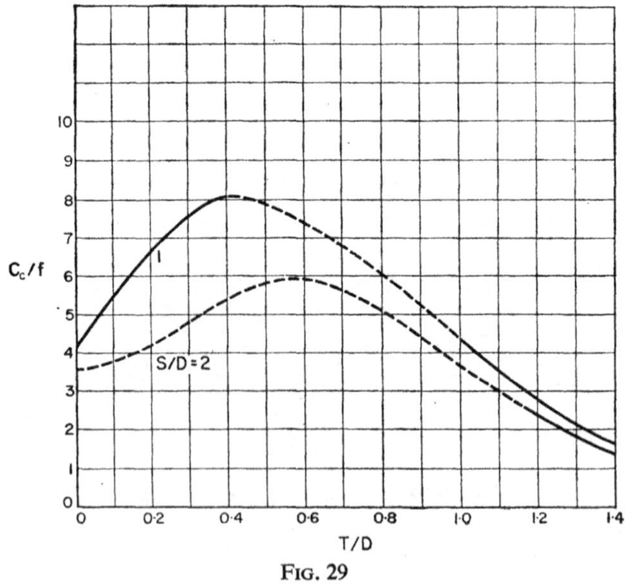

10
9
8
7
6
5
4
3
2
1
0
C_c/f
1
S/D=2
0
0·2
0·4
0·6
0·8
1·0
1·2
1·4
T/D

FIG. 29

The most useful results for unipotential electrostatic lenses stem from a development of the Glaser and Schiske analysis by Archard[18]. Some results replotted from his paper are shown in Figs. 28 and 29. The geometrical parameters were explained in Fig. 17. The dotted portions of the curves indicate that the object point is within the lens field and not therefore usable as an objective. Comparison with Fig. 17 shows that here again the minimum spherical aberration is achieved for the strongest lens. The minimum normalized C_s is seen to be somewhat higher than for the magnetic lens. In practice the comparison is even more in favour of the magnetic lens for electron microscope objectives as they can be made to have a shorter focal length. Finally, Fig. 29 shows the chromatic aberration, which is seen to have a maximum in the region where C_s is a minimum. The curve shown applies to a zero potential for the central electrode. It is of course possible greatly to enhance the chromatic aberration by using a negative central potential, and Möllenstedt[73] has used this arrangement as a velocity spectrometer.

5. CATHODE IMAGING AND ELECTRON MIRRORS

In our discussion of imaging, we have so far excluded situations where the electrons might reach a point where either the axial or the total velocity becomes momentarily zero. Since electrons start from a cathode, this situation is naturally encountered in the design of all electron emission systems, whether they are "electron guns" or some form of "image tube". There is little distinction in principle between these two cases, but they are sharply differentiated both in purpose and in the range of parameters which are involved. Thus the primary consideration for electrode guns is the formation of an electron beam having certain geometrical and current density specifications; in most applications a high current density is required. As a consequence, the effect of space charge is usually substantial. In the case of image converters, the primary source of electrons is photoelectric, and the emphasis is on producing an image of the cathode itself. Space charge is usually completely negligible. We will not be able to discuss electron guns in this work; electron guns for electron microscopes are discussed in some detail in Haines's book[74].

It is also possible to realize electron mirrors. At the moment of reflexion, the axial electron velocity is momentarily zero. Some aspects of the optics of mirrors are therefore very similar to the optics of cathode imaging. Mirrors are of importance in some situations where the reflexion of an electron beam is required as part of some other function, such as reflex klystrons and certain types of image converters. It is, however, becoming of increasing importance in another connexion. At the moment of reflexion, the electrons are very easily perturbed by small electric or magnetic fields. If one therefore arranges for the reflexion plane to be very close to some object whose surface one wishes to examine, then the electron beam can form an image of those surface characteristics which are associated with perturbing fields[(108)]. This technique has been used for examining magnetic domain patterns[(75)], but also to examine conductivity patterns[(76)]. In the latter case these were developed by the simple device of passing a small electric current through the object. Mirror microscopy can also reveal topographical details[(77)] by virtue of the perturbation of the applied mirror field. This perturbed field will naturally be proportional to the magnitude of the local "altitude" above the ideal mirror surface.

With so much of the rapidly developing technology of solid state integrated circuits being concerned with surface phenomena, the study of surfaces by electron mirror microscopy is becoming of increasing importance. An important limitation of the technique is set by the need for the surface to be examined to be relatively free from macroscopic roughness. In this respect the alternative technique—scanning microscopy[(78)]—is more widely applicable.

In this section we will be able to give only a very brief account of how the presence of a zero potential surface affects the analysis of electron optical devices.

5.1. Cathode imaging

The problem of imaging an object is to recombine in one point all the rays which traverse the corresponding object point in slightly different directions. If it were possible to devise an illuminating beam which is perfectly laminar, and which is partially absorbed (but not scattered) by the object, perfect images or rather perfectly sharp shadows of the object would be found at all subsequent points. Magnification could be achieved by using an appropriately divergent beam.

The laminarity of the beam of electrons emitted by a cathode is impaired only by the small thermal emission velocities. The high degree of laminarity is used directly in the emission microscope, and has led to the highest resolution limits so far achieved[79]—although the range of applications is limited by the need for the cathode to serve as the object. The laminarity depends on the ratio of the accelerating potential to the magnitude of the emission velocities. In the emission microscope high accelerating potentials are therefore employed, and for the attainment of the highest resolutions, the cathode is cooled in order to reduce the emission velocities.

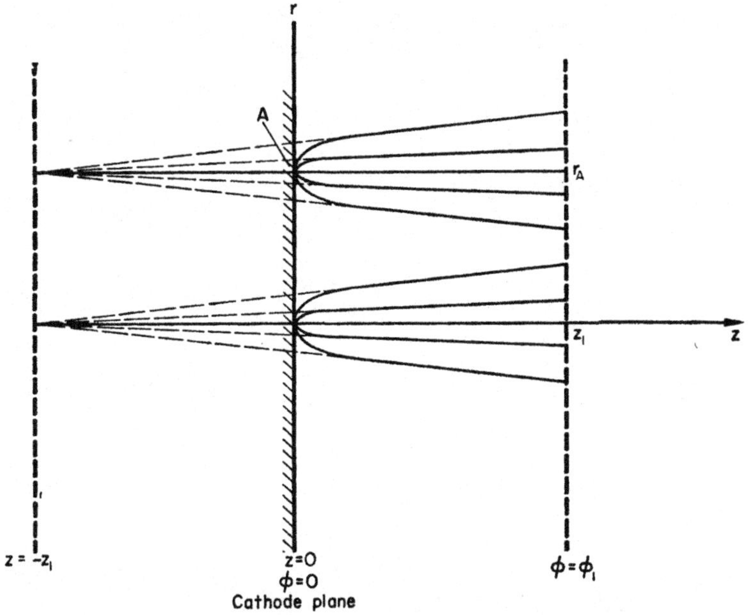

FIG. 30. Location of virtual object plane (radial fields neglected).

In the case of image tubes, where the accelerating potentials are relatively low, and where the emission velocities are generally high, the most probable velocity being typically slightly less than hf, where f is the frequency of the illuminating radiation—of the order of 1 V for visible radiation—sharp images will be obtained only at specific image planes. Nevertheless, some primitive but quite effective image converters have been built simply by placing a fluorescent screen at a positive potential, very close to the photocathode.

The analysis of cathode imaging in the general case is quite complex. Fortunately, it is possible to make an approximation, which is always valid sufficiently close to the cathode, and which greatly simplifies the situation. This is to neglect the radial electric field by comparison with the axial field. Since the cathode will be an equipotential, the lines of force in its immediate neighbourhood must be normal to the surface. We may consider the system shown in Fig. 30. Here ϕ_1 is an equipotential sufficiently close to the cathode to allow us to neglect its curvature, and also the curvature of the cathode itself if this is not in fact planar. We consider the electrons emitted at a typical point A on the cathode. As a result of the velocity distribution, a typical electron might have a radial velocity of $(2\eta\phi_T)^{\frac{1}{2}}$, where $\phi_T \sim kT/e$. In addition to this it will, of course, also have an axial velocity, but for the present purpose the effect of this will be neglected. Since the field is assumed to be purely in the z direction, the radial displacement of the thermal electron from its original position r_A will be Δ, where

$$\Delta = (2\eta\phi_T)^{\frac{1}{2}}\tau, \tag{5.1}$$

where τ is the transit time from $z = 0$ to $z = z_1$,

$$\tau = \int_0^{z_1} (2\eta\phi_1 z/z_1)^{-\frac{1}{2}}\,dz$$

$$= \frac{2z_1}{(2n\phi_1)^{\frac{1}{2}}}. \tag{5.2}$$

Hence

$$\Delta = 2z_1(\phi_T/\phi_1)^{\frac{1}{2}}. \tag{5.3}$$

The slope of the trajectory at $z = z_1$ is θ_1, where

$$\tan\theta_1 = (\phi_T/\phi_1)^{\frac{1}{2}}. \tag{5.4}$$

Thus the tangent to the trajectory at this point will intersect $r = r_A$ at

$$z = z_1 - \frac{\Delta}{\tan\theta_1} = -z_1. \tag{5.5}$$

Since this result is independent of ϕ_T, we can conclude that the virtual object is located z_1 behind the cathode.

The equipotential ϕ_1 simply delineates the region for which the approximation that the radial field is negligible is tenable. This would clearly apply also if ϕ_1 was produced by a diaphragm pierced by a small (i.e. $\ll z_1$) diameter hole. In this case there will be a sudden change in the slope of the trajectory given by r' in eq. (3.2) where ϕ' is the difference in the field just to the left and just to the right of the aperture. In the particular case where there is no field to the right of the aperture, this leads to an effective focal length of the aperture lens of $-4z_1$. Applying the lens formula to this case, and using eq. (5.5) for the effective object distance, we find for the image distance v:

$$\frac{1}{v} = -\frac{1}{2z_1} - \frac{1}{4z_1},$$

$$v = -\frac{4}{3} z_1. \tag{5.6}$$

In practice, the aperture in the diaphragm is rarely small enough to make eq. (5.6) a good approximation. For example, if the aperture diameter is equal to z_1 the virtual image is located only $0.19z_1$ behind the cathode[(80)]. In the case of electron guns, the situation is further complicated by the need to take space charge effects into account.

The result of eq. (5.5) is useful in giving first order answers to problems involving cathode lenses. For detailed knowledge of the trajectories, which might be required in a high quality image converter, the numerical techniques which would be adopted can, of course, be used right to the cathode, without the need for any specific assumptions[(81)].

The use of magnetic fields in cathode imaging is also of considerable importance. A very simple and much used converter, shown in Fig. 31, is based on the use of a strong magnetic field so that the trajectories remain in close proximity to the magnetic field lines. An individual electron will describe a circular motion in the transverse plane with radius proportional to $\phi_T^{1/2}$ but a period which is independent of ϕ_T. Since the axial velocities of all electrons are very nearly identical, electrons of different thermal velocities are brought to a focus in the same plane. This last statement ignores the slight scatter in axial velocities arising from the axial components of the emission velocity. This is generally a small effect. However, it is worth noting that it is cumulative; thus if multiple images are formed in a long tube, the successive images will be progressively less perfect.

The effect is exacerbated if the magnetic field is increasing with distance.[82].

The use of magnetic fields in cathode lenses is particularly important for transmission electron multipliers[83]. Here the task of the magnetic field is to image each progressive dynode on the next. The great advantage which springs from the use of magnetic fields instead of pure electrostatic fields lies in the fact that it allows the focusing to be adjusted independently of the brilliance and, in some cases, of the magnification.

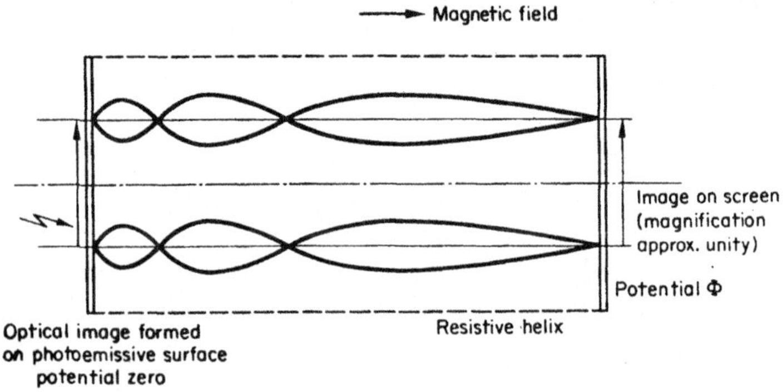

FIG. 31. Simple image converter. Electric field approximately uniform. The magnetic field is sufficiently strong to prevent large excursion from a magnetic line of force. In the diagram the magnetic field is adjusted so that the third cathode image is utilized.

5.2. Electron mirrors

Electron mirrors can be formed by any circularly symmetric structure which contains within it a zero equipotential crossing the axis. A typical example is the three-electrode lens, and typical trajectories are shown in Fig. 32. The electron optical properties of such mirrors have been investigated in great detail. For some time, the fact that the spherical aberration of such a mirror is opposite in sign to that of all lenses led to the hope that they represented a possible path to the improvement of electron microscopy resolution. Practical considerations have led to the fading of such hopes—not the least of the troubles being due to the embarrassing fact that the reflected electron beam is in the same region as the incident beam, unless the system is tilted, a solution which tends to lead to further aberrations. Mirrors of the form shown in Fig. 32 are used in some image converters. However, by far the most important application of the mirror principle is in obtaining information about a physical surface, i.e. when the mirroring surface itself represents the "object". In addition to the cases already noted, one should mention that of the semiconductor image converter, in which the incident radiation causes the semiconductor film to be charged in sympathy with the intensity of the light. This potential distribution can then be revealed

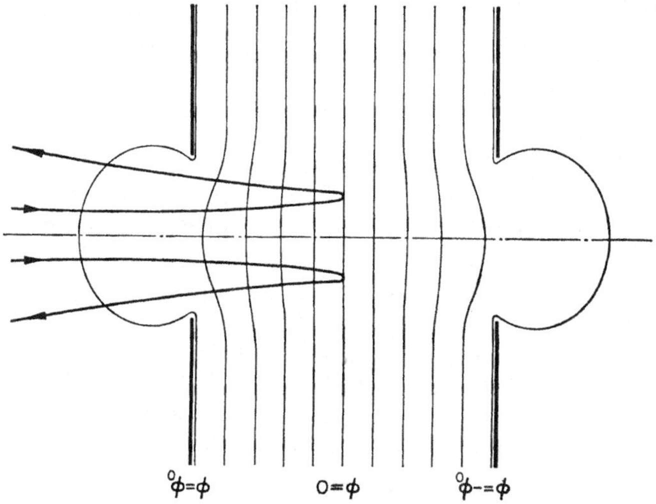

FIG. 32. Ray path in a non-uniform field mirror (after Zworykin *et al.*, ref. 80). (Direct reproduction of Fig. 13.32, p. 462, Zworykin *et al.*)

by making the rest of the semiconductor surface act as a mirror in a suitable electron optical system. This type of device has been developed for infra-red image converters[(84)], for wavelengths at which normal photoemissive materials have a very low quantum efficiency.

For practical purposes, the primary interest lies with simple planar electron mirrors. As we will see, these are not quite as simple as optical planar mirrors. In particular, they are not free from spherical aberration. The theory is very similar to that for the emission system of Fig. 30.

Suppose that we illuminate an object located at $z = z_0$, at a potential ϕ_1, with a beam coming from the $+z$ direction. All the electrons will be reflected before they reach the plane $z=0$, which represents the zero equipotential surface, the value of z where reflexion takes place being larger for those electrons which have large radial velocities. Consider a particular electron leaving the object plane inclined at an angle θ to the axis. When it enters the uniform retarding field region its trajectory will be parabolic, and it will return to the object plane at a radius r_1, once again inclined at an angle θ. It is readily shown that the radial displacement, $r_1 - r_0$, is given by

$$r_1 - r_0 = 2z_0 \sin 2\theta. \tag{5.7}$$

The tangent to the emerging ray will intersect $r = r_0$ at $z = z_i$ where

$$z_0 - z_i = 4z_0 \cos^2 \theta. \tag{5.8}$$

Thus for paraxial rays

$$z_i = -3z_0. \tag{5.9}$$

Thus the image is located three times the object distance behind the "mirror" plane and is not free from aberration [i.e. the $\cos^2\theta$ factor in eq. (5.8)]. The reason for this behaviour is that the reflexion takes place gradually in the $0 < z < z_0$ range. In principle, one could reproduce the behaviour of an optical plane mirror by placing the object in field free space, and using a fine grid to confine the reflexion region to a length which is small compared with the object distance.

If we are using the mirror for surface examination, the object is, of course, located at (or very close to) the mirroring plane. Under these conditions, the function of the incident beam is purely to illuminate the object, and the optical behaviour of the system becomes that of the simple emission system of Fig. 30. Thus the virtual object is located at a distance $-z_1$ behind the surface which is being examined. In practice the $z = z_1$ plane is, of course, defined only in the sense that it is sufficiently close to the $z = 0$ plane to allow one to regard it as an equipotential, and yet sufficiently far away to justify the paraxial assumption that $r' \ll 1$. Normal paraxial theory is then applicable to the region $z > z_1$.

In the more general case, for a mirror such as that shown in Fig. 32, it is possible to derive a "thin mirror" approximation for the effective focal length in terms of the axial distribution of the potential, an expression which survives the hazards of the infinite value of r' at the reflexion point. However, for computational purposes the direct attack is as fast, and avoids any difficulties at the reflexion point.

There are various possible approaches: perhaps the simplest is to define an equipotential surface ϕ_1, and use plane mirror theory between this and the zero equipotential, resorting to normal paraxial theory for all larger values of ϕ.

6. THERMAL VELOCITIES

In many applications an electron beam is used as a form of "pointer"; in such cases one is usually striving for the highest possible resolution. This means that we are concerned with focusing an electron beam to the smallest possible spot size. The ordinary cathode ray oscilloscope is an example in this class. Others are the scanning microscope(85), the X-ray microprobe(86), and electron beam tools of the kind used in "writing" on to photoresist in semiconductor microcircuit applications(87).

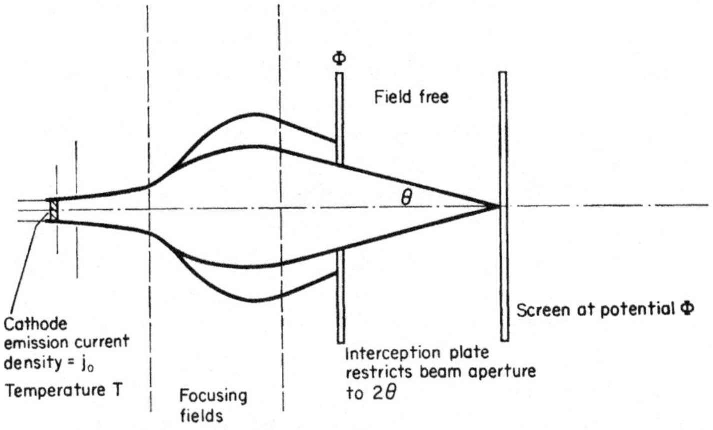

FIG. 33. Arrangement for thermal velocity calculations. The beam is not necessarily "focused" on the screen.

Whilst the spot size determines the attainable resolution, one is also concerned with the magnitude of the current within a spot, since this will control other vital performance characteristics. For example, the scanning speed in the scanning microscope will be directly proportional to the current in the beam. In such cases one is therefore primarily concerned with the problem of obtaining the maximum possible current density within a small spot diameter. A frequently adopted arrangement is to focus the beam to a small spot using a converging lens, Fig. 33. The question then arises as to what factors will limit the attainable current density. One can readily identify four such factors:

(i) finite electron wavelength;
(ii) lens aberrations;
(iii) space charge;
(iv) finite thermal emission velocities.

The finite electron wavelength will rarely make itself felt directly, since this wavelength is 10 Å at only 1.5 V. If this were the only limitation, it would be possible to obtain spot diameters well below 1 Å at quite moderate voltages such as 1 kV. In practice, it is other factors which under most circumstances represent the effective bar to improvements.

The spherical aberration of the lens will clearly be damaging—the damage depending on the cube of the aperture angle, θ^3. At first sight one might conclude that, to reduce its effect, all we need do is place the focus far from the lens, and thus work with a small value of θ, at least until we reach the point where electron diffraction becomes serious. However, as we shall see, this would exacerbate the effect of the thermal velocities.

Space charge can represent a real limitation. A simple application of Gauss's law shows that the radial field at the edge of the beam, due to space charge, is inversely proportional to the beam radius. The divergent forces near the beam minimum can therefore dominate the situation. This is indeed the case for "heavy current" beams, for which see Part 5. In the case of microscopes and oscilloscopes, it turns out that the space charge effects are generally negligible. However, this is not invariably so, and particularly in the case of electron beam "tools" for microminiature circuit fabrication, the effect may have to be considered. For the present we will, however, assume that it can be neglected.

It is generally a combination of the thermal velocities and the lens aberrations which govern what can be achieved in practice. It may at first sight appear surprising that the thermal velocities, which have an energy which is typically less than 0.1 eV, should have such an important bearing on the behaviour of the beam, even after it has been accelerated to a voltage of perhaps 20 kV. It must, however, be appreciated that, at the focus, the diameter of the beam will be far less than its value at the cathode—i.e. the beam has been severely compressed. It can be shown quite simply that this compression leads to an increase in the r.m.s. value of the random component of the transverse velocities[88], or, speaking somewhat pictorially, that the transverse electron temperature has been increased. It is in fact increased by M^{-2}, where M is the linear magnification. If, for example, we are seeking to produce a spot with an area of 1 μ^2, from a cathode having an area of 0.1 mm^2, the effective transverse temperature has been increased from its value at the cathode of, say, 1000°K to 10^8 °K, and typical transverse energies may be within a factor of 10 of the forward energy.

Before discussing the analysis of this situation, it is of interest to appreciate the nature of the loss which is incurred as a result of the presence of the thermal velocities. If the lens in Fig. 33 is assumed to be perfectly free from aberration, it should be capable of producing a perfect image of the cathode reduced by a factor as large as we may desire. However, beyond a certain point, the current density in this image will not continue to increase. Where has the missing current gone? The answer is that even though the lens is as free from aberration as one can conceive, some of the current has none the less failed to reach the image—an "ideal" lens, which can accept trajectories with arbitrary inclination to the axis, and which recombines them all in a perfect image of reduced dimensions cannot exist.

The law which governs the limiting attainable density was first discovered by Langmuir[89], and later more general proofs were given by Gabor[90] and by Pierce[91]. In the next section we will outline the proof given by Pierce[92]. This proof is applicable to a situation when the focus lies in field-free space. It turns out that somewhat different considerations apply if it is located in a strong magnetic field. This point will be briefly considered in section 6.2. The relevance of these results to the problems of electron beam deflexion will be briefly noted in the final section.

6.1. The thermal velocity limit—focus in field-free space

The starting point in the proof is Liouville's theorem: the flow of an ensemble of non-interacting particles is such that the local density of the particles in phase space (velocity times configuration space†)

† The more usual definition is momentum–configuration space. It turns out that here (even in the presence of a magnetic field) equivalent results are obtained in velocity–configuration space.

remains constant. If the density at one instant in time is N_0 and we observe the density in the vicinity of the same particle at a later instant is N_1, then

$$N_0 = N_1. \tag{6.1}$$

This result is simply a conservation statement. The only assumption is that the interaction between the particles can be neglected. The actual charge contained in an infinitesimal volume of phase space is therefore $\mathrm{d}Q$, where

$$\mathrm{d}Q = N \,.\, \mathrm{d}v_x \mathrm{d}v_y \mathrm{d}v_z \mathrm{d}x \mathrm{d}y \mathrm{d}z. \tag{6.2}$$

If we know the velocity distribution of the electrons at the cathode, we have complete knowledge of the function N at one plane in the beam—the cathode plane—and are then in a position to deduce its form at other points in the beam by means of eq. (6.1). The distribution function at the surface of a thermionic cathode will be very nearly half-Maxwellian, so that N_0, the form of N at the cathode, will be

$$N_0 = K \exp\left[-\frac{m}{2kT}(v_{x0}^2 + v_{y0}^2 + v_{z0}^2)\right] = K \exp\left(-\frac{m}{2kT} v_0^2\right),$$

where

$$v_0^2 = v_{x0}^2 + v_{y0}^2 + v_{z0}^2.$$

We now wish to deduce a general function $N(v, \Phi)$ which will certainly depend on the velocities v and the potential field ϕ. Writing the energy integral [eq. (2.12)] in Cartesian coordinates

$$v^2 = 2\eta\Phi + v_0^2. \tag{6.3}$$

This suggests that the appropriate form for $N(v, \Phi)$ is

$$N(v, \Phi) = K \exp\left[-\frac{m}{2kT}(v^2 - 2\eta\Phi)\right], \tag{6.4}$$

a function which by virtue of eq. (6.3) leads to a constant density in phase space, and moreover tends to the known distribution function at the cathode where $\phi = 0$. It is interesting to note that the magnetic field does not enter into this expression. Whilst it will undoubtedly influence the density in velocity space, and also the density in configuration, these influences will compensate such that the density in phase space remains unaffected.

We will wish to concentrate our interest on the current flowing in a particular, the $+z$, direction. Noting that

$$j_z = \rho \, . \, v_z$$

and that ρ, the charge density in configuration space, is simply $Q/\mathrm{d}x\mathrm{d}y\mathrm{d}z$, we have from eqs. (6.2) and (6.4)

$$\mathrm{d}j_z = K v_z \, . \exp\left[-\frac{m}{2kT}(v^2 - 2\eta\Phi)\right] \mathrm{d}v_x \mathrm{d}v_y \mathrm{d}v_z. \tag{6.5}$$

It remains to integrate this expression over the velocities. We must therefore choose the appropriate limits for each component of the velocity, a choice which is equivalent to stating which classes of electrons are able to reach the point of interest. The only place where this choice is at once clear is at the cathode, where v_x and v_y will range between plus and minus ∞, and v_z between 0 and ∞. Carrying out this integration, we are then able to express the constant K in terms of j_0, the current density at the cathode:

$$j = \frac{1}{\pi}\left(\frac{m}{kT}\right)^2 j_0 \, \varepsilon^{e\Phi/kT} \int \varepsilon^{-mv_x{}^2/2kT} \mathrm{d}v_x \int \varepsilon^{-mv_y{}^2/2kT} \mathrm{d}v_y \int \varepsilon^{-mv_z{}^2/2kT} v_z \, \mathrm{d}v_z. \tag{6.6}$$

In order to carry out this integration for the circular symmetric arrangement of Fig. 33, we transform to spherical coordinates in the velocity space, v and α, where α is measured from the z axis. Now since the minimum value of v at the cathode v_0 is zero, we know at once that the minimum value at the point of interest is $(2\eta\Phi)^{\frac{1}{2}}$. The maximum value is infinity since we start from a Maxwellian distribution. The minimum value of α is, of course, zero, and the maximum value will be set by some limiting aperture which for convenience can be regarded as located outside the lens field, and which subtends an angle θ at the point of interest. With these limits, the integration of eq. (6.6) yields the following remarkable simple result:

$$j_m = j_0\left(1 + \frac{e\Phi}{kT}\right)\sin^2\theta, \tag{6.7}$$

where we have written j_m to emphasize that the expression of eq. (6.7) yields a maximum value for the attainable current density. It is a result which is applicable at an "image" of, for example, the cathode, at a crossover point or at any other arbitrary location. It depends only on the potential *at* the point of interest, and in no way on the potential or magnetic field distribution between the cathode and this point. The only restriction is that it is possible to define the maximum angle θ of arrival of the electrons in terms of the location of some physical aperture.

This generality is, of course, a consequence of the fact that we have used very little physics in deriving the result. We have appealed only to the conservation of particles (Liouville's theorem) and to the conservation of energy. It is therefore clear that the result would apply equally to a swarm of particles in a gravitational field, or to a swarm of photons. Indeed, eq. (6.7) is the direct analogue of the law of conservation of brightness in optics. The big practical difference between optics and electron optics is that in the latter one can increase the refractive index ($\phi^{1/2}$) by orders of magnitude. Thus, for example, the effective brightness at 10 kV is five orders of magnitude greater than that at the source—the cathode. In some applications it is the maximum attainable power density which is of primary interest. This is seen to be proportional to ϕ^2.

In asking how closely one is able to approach to the limit given by eq. (6.7) we must bear two different considerations in mind. Suppose that we wish to form a reduced image of the cathode by a focusing system, which for the moment we regard as free from aberration. This case, which has been analysed in detail by Pierce[(92)], shows that one can indeed approach the limit as closely as we might wish, but at the cost of wasting progressively larger portions of the emitted current. As discussed above, the focusing system, although free from "aberrations", cannot in fact recombine all the rays from one point in the object at the corresponding point in the image. For example, Pierce finds that in order to approach within 70 per cent of the j_m, it is necessary to waste about half of the emitted current.

This, then, is the basic engineering bargain which is involved. Over and above this, aberrations of the focusing system will lead to further reductions in the attainable current density.

6.2. The thermal velocity limit—focus in a magnetic field

At first sight it appears surprising that the local magnetic field does not enter explicitly into the limiting expression of eq. (6.7). In fact it does have an influence on the result, but only in that it influences the effective value of θ. If the point of interest lies in the magnetic field, the trajectories will be curved, and θ will no longer be capable of definition in terms of a physical aperture. This situation is of considerable interest, because for some applications a type of focusing system, which has frequently been used in heavy current (i.e. space charge dominated) beams, can with advantage be used for "low current" applications. Figure 34 shows the main features of the system considered. The beam is subject to a magnetic field, which increases monotonically from its value at the cathode, thus gradually

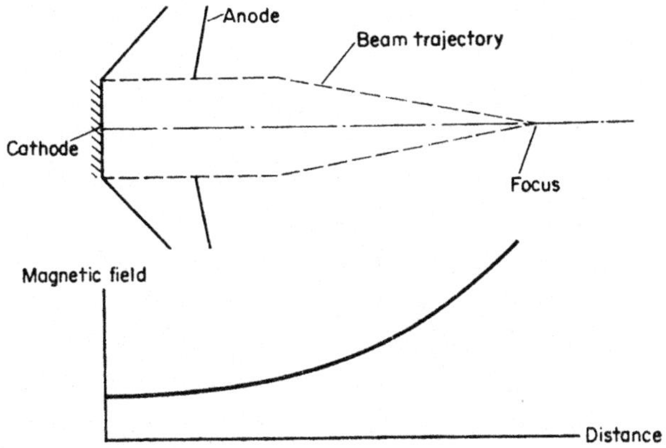

FIG. 34. Confined flow magnetic compressions system.

compressing the beam. The determination of the attainable current density in this case represents a complex problem, but it becomes tractable for the particular case when the field, even at the point where it takes its lowest value, the cathode, is assumed to be extremely large. Under these conditions one can at once surmise that the limiting current density should be that of eq. (6.7) with $\theta = \pi/2$, since, in a large magnetic field, the electrons can approach the point of interest at right-angles to the axis. The following analysis confirms this conclusion.

Let the magnetic field at the cathode be B_0, and at the point z be $R.B_0$. Our starting point is again eq. (6.6). However, here we will change to cylindrical rather than spherical coordinates in velocity space, as we need to consider the axial velocity separately. We therefore write

$$v_x^2 + v_y^2 = v_t^2 \tag{6.8}$$

and eq. (6.6) becomes

$$= 2\left(\frac{m}{kT}\right)^2 j_0 \, \varepsilon^{\Phi/kT} \int v_z \, \varepsilon^{-mv_z^2/2kT} \, \mathrm{d}v_z \int v_t \, \varepsilon^{-mv_t^2/2kT} \, \mathrm{d}v_t. \tag{6.9}$$

Now as the compression of the beam proceeds, the transverse velocities will increase, at the expense of the longitudinal velocities.

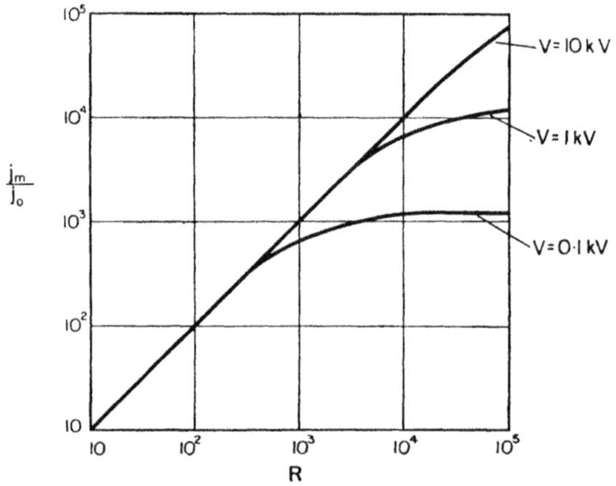

FIG. 35. Limiting current densities for large magnetic fields.

If the compression is sufficiently slow the theorem of adiabatic invariance[(94)] leads to an increase of transverse energy which is directly proportional to R,

$$v_t^2 = R \, . \, v_{t0}^2. \tag{6.10}$$

We can use this in the energy integral, and hence find the maximum value that v_t can have if v_z is to remain positive. This expression, which is a function of R, then provides the top limit of the v_t integral in eq. (6.9). The appropriate limits for v_z are not easily established directly, but those for v_{z0} are, of course, known. It is therefore easiest to proceed by changing the variable from v_z to v_{z0} in eq. (6.8).

The integrals can then be carried out, and lead to the following result

$$j_m = j_0\left[R-(R-1)\exp\left(-\frac{e\phi}{kT(R-1)}\right)\right]. \tag{6.11}$$

It is readily seen that for $R \gg 1$, eq. (6.11) becomes

$$j_m = j_0(1+e\phi/kT) \tag{6.12}$$

in accordance with the considerations advanced earlier. The general expression has been computed in Fig. 35 for a cathode working at 1000°C, and for a number of different beam potentials. The way in which the situation is changed when the infinite magnetic field approximation is removed can be predicted to some extent from simple semiquantitative considerations[(93)]. It would appear that very high current densities could be achieved by this method; moreover, in contrast to the situation when one attempts to work with large values of θ in the normal lens optics, there are no lenses close to the high current density point. There is therefore much more space to accommodate deflexion plates or coils.

7. ELECTRON BEAM DEFLEXION

In the class of applications in which the electron beam is used as a "pointer", the ease with which the beam can be deflected is a primary consideration. Taking the cathode ray tube (c.r.t.) as an example, we have a device which is basically very simple, which can readily display (or memorize) in excess of 10^6 points. Any one of these points can be picked out (i.e. illuminated) at random, with a delay time of only a few nanoseconds. This represents a performance which, at present, solid state display panels cannot begin to approach; in any event the associated circuitry would be very complex. In a technology in which digital techniques are rapidly displacing analogue techniques, the deflexion capability of the electron beam in a c.r.t. is likely to ensure the survival of this analogue device for the foreseeable future.

In those applications where the electron beam is used as some sort of a processing tool, the deflexion capability is again of critical importance. In many of these applications, laser beams are possible alternatives—but start from the initial disadvantage that it is vastly more difficult to effect laser beam deflexion at high speeds.

Electron beams can be deflected by electric and magnetic fields; both techniques are in current use. Usually, these deflecting fields are at least approximately "two-dimensional", so that they vary along the z axis, and with x, but not with y. The first order theory for this case is relatively simple, and follows very much the same lines as first order lens theory. If this restriction on the field is removed, and if aberration effects are of importance, a theory can still be developed in terms of the axial values of the potentials or fields, and their derivatives in different directions[95, 96]. However, this approach usually falls into that class of theoretical solution which is no closer to a numerical result than are the starting equations—which can be the direct basis of a computation. References to such direct computer calculations will be made in the following. We will confine ourselves, in section 7.1, to giving the elementary results for the simplest cases.

In many applications it is not so much the actual deflexion which matters as the deflexion measured in spot diameters, for it is this which will determine the attainable resolution. The problem of deflexion is then intimately bound up with the spot size as determined by thermal velocity effects, a topic which will be taken up in section 7.2.

It is a property common to all deflexion fields that they possess focusing as well as deflexion properties. An electron beam, which is properly focused before deflexion, will therefore be defocused to an extent which depends on the angle of deflexion. A brief account of the deflexion aberration problem—and methods of cure—will be given in section 7.4.

7.1. Deflexion by one-dimensional fields

In this section we will consider the deflexion produced by a field $E(z)$ or $B(z)$ which are not functions of either x or y. The system with which we are concerned is shown in Fig. 36, specifically for the electrostatic deflexion case. For magnetic deflexion, we will consider pole pieces in a plane at right-angles to that of the electrostatic deflector plates.

In the electrostatic case, we will assume that in the region $0<z<a$ the deflecting field is $E_d(z) = \phi_d/b(z)$ where $b(z)$ is the plate separation at z. Thus we neglect "fringing fields" and moreover imply that $b'(z) \ll 1$. The transverse equation of motion then becomes

$$\frac{d^2 y}{dt^2} = \eta E(z) = \frac{\eta \Phi_d}{b(z)}. \tag{7.1}$$

For the moment we will concern ourselves solely with the central trajectory, and make the assumption, which is directly analogous to the paraxial assumption in first order lens theory, that the axial velocity v_0 is constant. In doing this we are in fact neglecting two distinct effects: firstly, the reduction in longitudinal energy which ust arise to provide the energy in the transverse motion, and

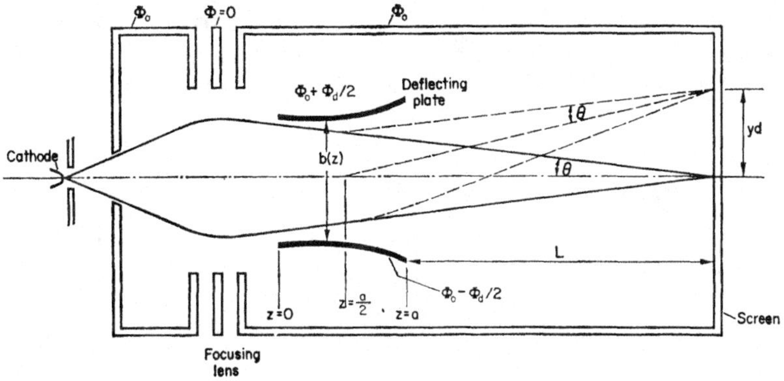

FIG. 36. Basic arrangement for electron beam deflexion. For magnetic deflexion the magnetic field is normal to the plane of the paper.

secondly, the change in total energy of the electron due to its departure from the ϕ_0 equipotential. With this assumption

$$\frac{d^2}{dt^2} = v_0^2 \frac{d^2}{dz^2}$$

and

$$y'' = \frac{\eta}{v_0^2} \frac{\Phi_d}{b(z)} = \frac{\Phi_d}{\Phi_0} b^{-1}(z). \tag{7.2}$$

Integrating twice,

$$y = \frac{\Phi_d}{\Phi_0} \int_0^z dz \int_0^z b^{-1}(\xi)\, d\xi, \qquad 0 < z < a.$$

For $z > a$, y' is constant and the final deflexion is therefore y_{de}, where

$$y_{de} = \frac{\Phi_d}{\Phi_0} \left[\int_0^a dz \int_0^z b^{-1}(\xi)\, d\xi + L \int_0^a b^{-1}(\xi)\, d\xi \right]. \tag{7.3}$$

For the special case of parallel plates, eq. (7.3) reduces to the well-known expression

$$y_{\text{de}} = \frac{\Phi_d}{\Phi_0}\frac{a}{b}\left[L+\frac{a}{2}\right]. \qquad (7.4)$$

One may interpret this in terms of an "equivalent thin deflector" located at the centre of the deflecting plates.

The results for the magnetic deflexion case are entirely similar; however, to avoid the introduction of magnetomotive forces, we prefer to specify the field $B(z)$ in terms of its value at $z = 0$, so that

$$B(z) = B(0)\,.\,b(0)\,.\,b^{-1}(z). \qquad (7.5)$$

This leads to the following expression for y_{dm}:

$$y_{\text{dm}} = \left(\frac{\eta}{\Phi_0}\right)^{\frac{1}{2}} B(0)\,.\,b(0)\left[\int_0^a \mathrm{d}z \int_0^z b^{-1}(\xi)\,\mathrm{d}\xi + L\int_0^a b^{-1}(\xi)\,\mathrm{d}\xi\right] \qquad (7.6)$$

and for parallel pole pieces,

$$y_{\text{dm}} = B\left(\frac{\eta}{\Phi_0}\right)^{\frac{1}{2}} a\left(L+\frac{a}{2}\right). \qquad (7.7)$$

Once again we find the equivalent thin deflector is located half-way along the length of the deflecting field.

One of the aims in designing deflecting plates is to minimize the (reactive) control energy required to deflect the beam. Since the energy stored in the deflexion region is proportional to the square of the field, whilst, as appears in eqs. (7.4) and (7.7), the deflexion depends linearly on both the field and the length of the deflexion region, it is clearly advantageous to make the deflexion region as long as possible. There are two considerations which limit attempts at improvement in sensitivity by this means: the plates must be designed so that even for the largest deflexions no interception occurs. Thus $b(z)$ must increase with z, and the deflexion will increase more slowly than linearly with a. The second consideration, which applies to the high frequency c.r.t., is the need to ensure that the transit time through the plates is substantially less than one period of the highest frequency voltage which is to be observed.

To reduce the stored energy, the plates should clearly be designed so as to hug the beam profile in its maximum deflexion position;

any energy stored in regions which are never traversed by the beam clearly represents an unnecessary demand on the deflexion source. It is not difficult to calculate the appropriate shapes of the deflecting plates (or pole pieces). Figure 37 shows the form, but in practice it is more usual to approximate this by a parallel followed by a flared section. It can be shown that the penalty incurred by using this approximation is very slight[(97)]. In a practical design, in addition to the requirement for minimizing stored energy, one is also concerned with defocusing effects (section 7.4) and, of course, practical problems—such as winding coils around the glasswork of a c.r.t.

One direction in which one might hope to reduce the deflexion energy is to effect the deflexion at a relatively low potential ϕ_0, and then subsequently to accelerate the beam to the required final energy. The considerations governing "post-acceleration" techniques of this kind are intimately tied up with the thermal velocity effects. A brief discussion will therefore be deferred to section 7.3.

7.2. Thermal velocity effects

The minimum observable deflexion of a beam is of the order of the diameter of the spot formed on the target surface. For many applications interest is therefore centred on the deflexion field required to effect this minimum deflexion. Consider again the system of Fig. 36. Since the aperture of the beam at the target surface is 2θ, we know that at no point in the spot can the current density exceed the value of j_m given by eq. (6.7). In practice, if the beam is focused to its narrowest diameter, the actual current distribution is likely to be approximately Gaussian, so that only near the centre can the current density approach j_m. Nevertheless, if we define a diameter d_m by the equation

$$\frac{\pi}{4} d_m^2 j_m = I, \tag{7.8}$$

where I is the total beam current, d_m will in most cases serve as a fair approximation to the actual beam diameter. If, for example, the beam diameter is defined by the points where the current density falls to half the central value, it will exceed d_m by a factor which might typically be of the order of two. For our present purpose it is even more important that, expressing d_m in terms of j_m, as given in eq. (6.7),

$$d_m = \frac{2}{\sin\theta}\left(\frac{I}{\pi j_0}\right)^{\frac{1}{2}}\left(1+\frac{e\Phi_0}{kT}\right)^{-\frac{1}{2}}, \tag{7.9}$$

the predicted functional dependence on θ and ϕ_0 is generally well maintained.

In practice the beam diameter at the beginning of the deflexion region d_0 will be determined by a number of considerations, such as the available space in the neck of a c.r.t. If we assume this to be fixed, and further than $L \gg a$, and $L \gg d_0$, we find from eqs. (7.4) and (7.9) that the deflecting potential ϕ_d required to effect a deflexion $y_d = d_m$ is

$$\Phi_d = 2\left(\frac{kT}{\pi e}\right)^{\frac{1}{2}} \frac{b}{d_0 a}\left(\frac{I}{j_0}\right)^{\frac{1}{2}} \Phi_0^{\frac{1}{2}}. \tag{7.10}$$

The corresponding expression for magnetic deflexion is

$$B_d = \frac{2}{n^{\frac{1}{2}} d_0 a \pi^{\frac{1}{2}}}\left(\frac{I}{j_0}\right)^{\frac{1}{2}}. \tag{7.11}$$

Thus, with these approximations, the required deflexion field is independent of L, in both cases; one cannot increase the total amount of information displayed simply by increasing the length of the pointer.

The dependence on ϕ is also of great interest; whilst for electric deflexion the deflexion capability improves as $\phi^{1/2}$, for magnetic deflexion it is totally independent of ϕ. This result will be seen to bear on the advantage to be obtained by resorting to post-deflexion acceleration.

7.3. Post-deflexion acceleration (p.d.a.)

The deflexion produced by an electrostatic field is, as seen in (7.3), to be proportional to ϕ_0 and $\phi_0^{1/2}$ for the magnetic deflector. This suggests that one can reduce the demands made on the deflexion system, by operating this at a lower potential than the screen[98, 99]. The "post-deflexion acceleration" can be effected in a number of different ways, some of which are shown in Fig. 37. The form of the axial potential distribution is indicated for each case. In case (a), the final acceleration takes place very close to the screen, the potential in the deflexion being maintained constant right up to the mesh which is placed just in front of the screen. This design involves a number of technical difficulties, notably that of avoiding loss of contrast due to secondary emission, and is therefore not widely used; however, we will find it of value as a basis for comparison with other types of p.d.a. In (b) the acceleration takes place gradually by

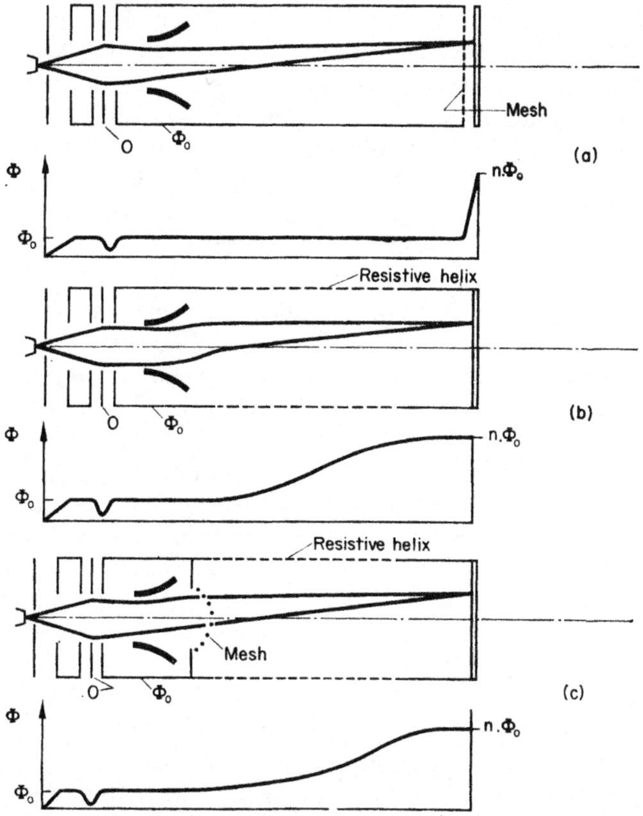

FIG. 37. Post-deflexion acceleration. (a) Final acceleration very close to screen. (b) Progressive acceleration between deflector and screen, by means of a resistive helix. (c) Progressive acceleration between mesh and screen by means of a resistive helix. The potential distribution in the acceleration region depends on the design of the helix and on the shape of the mesh.

means of a number of electrodes connected to suitable potentials. A very common technique is to paint a resistive helix on the vacuum envelope and developing the required potential distribution by applying a potential between its ends. In (c) the acceleration again takes place gradually, but here a mesh is placed at the beginning of the acceleration region.

Before discussing the relative merits of these different approaches to p.d.a., let us consider the effect of any of them on the "resolution sensitivity"—i.e. the field required for a deflexion equal to one spot diameter. Let us consider specifically the case of Fig. 37a. Since the final acceleration takes place in a very short distance, it is at once clear that neither the deflexion nor the spot size can be significantly affected. We can then at once conclude that the resolution sensitivity for electrostatic deflexion will remain precisely that given in eq. (7.10). Thus if the screen potential is $n\phi_0$, we have achieved an improvement by a factor $n^{1/2}$. Similarly, from eq. (7.11) we see at once that there is no improvement whatsoever in the case of magnetic deflexion.

We have reached these conclusions by reference to a particular potential distribution. Is it possible that by the use of a modified distribution one might achieve a greater measure of improvement? The answer, which was first given by Pierce[(100)], is a definite no. The form of the potential distribution between the deflecting plates and the screen is not relevant. For details of the proof the reader is referred to Pierce's book[(100)]. However, the ingenious argument used is at the same time very simple, and one can express the essence in a very few words: the thermal velocity distribution in the beam is manifested in that, at any point within it, the electrons passing through it form a cone of semi-angle α. If we focus the beam to the smallest possible spot, those electrons which impinge on the edge of the spot must have been those which at all other points in the beam—including the deflexion region—made the largest angle with respect to the central trajectory in each of the cones. Thus if the deflecting fields are such that all trajectories are rotated through the same angle α, all the electrons impinging on the screen will just miss the area previously covered by the spot—the beam will have been moved by one spot diameter. Then, the deflecting fields required to move the beam by one spot diameter are determined entirely by the conditions *within the region* occupied by the deflexion fields. The potential distribution thereafter (irrespective of whether it obeys Laplace's equation) cannot affect the resolution sensitivity. Alternatively, we may conclude that if the p.d.a. fields change the absolute deflexion achieved, they will change the spot diameter in the same proportion.

The improvement of $n^{1/2}$ in resolution sensitivity achieved for electrostatic deflexion is in any event worth striving for. However, as already discussed, p.d.a. certainly provides a greater absolute sensitivity, which can otherwise be achieved only by making the tube inordinately long. The form of the potential distribution naturally does affect the deflexion sensitivity attained. In the case of Fig. 37a we have seen that the full improvement by the factor n (screen/deflexion potential) is realized. It is easily seen that for the case of Fig. 32b the improvement achieved must be less; the net effect of the potential distribution is to reduce the deflexion. For a monotonic increase in potential this follows at once from the fact that Laplace's equation implies that E_r is negative if E_z is positive, a conclusion which is readily extended into the general case.†

The deflexion sensitivity achieved in p.d.a. systems of the type shown in Fig. 37b will therefore always fall short of the factor n achieved for electrostatic deflexion ($n^{1/2}$). An approach which overcomes this problem has recently been successfully implemented[(101)] (Fig. 37c). By the interposition of a mesh, the restraint imposed by Laplace's equation is relaxed. By reference to Fig. 38a it is readily seen that in this case the full improvement in deflexion sensitivity can be realized: the spherical mesh, maintained at approximately ϕ_0, is centred on the apparent centre of deflexion of the plates (i.e. at $z = a/2$), and the field between it and the spherical screen is radial. Here all deflected rays will impinge normally on to the mesh, and thereafter suffer no further change in direction—the trajectory being parallel to the electric field.

In practice this system would be modified in a number of different ways, some of which are indicated in Fig. 38b. It is indeed possible to obtain a deflexion sensitivity which gives a gain which is actually greater than n. The system shown in Fig. 38b is used in a commercial c.r.t.[(102, 103)]. Here the centre of deflexion is located beyond the centre of curvature of the mesh. As a result the final deflexion is further enhanced. A good approximation to a purely radial field is achieved immediately beyond the mesh by a suitably designed resistive helix. However, the screen itself is nearly flat, and no attempt to maintain the radial distribution is made beyond a certain point.

† In fact one can use the same proof as that used to establish the convergence of all lenses having zero field on either side [see eq. (2.50)]. The fact that here the field at the screen may be non-zero cannot matter, as one could always add a very short region in which the field decreases to zero. This can have no significant effect on the point of impact—even though it can, of course, change the angle of impact.

Since by that time the beam has been accelerated to a high velocity, fields transverse to the trajectories lead to only very small reductions in the total deflexion.

Here we have an application of a mesh lens—one of the first means

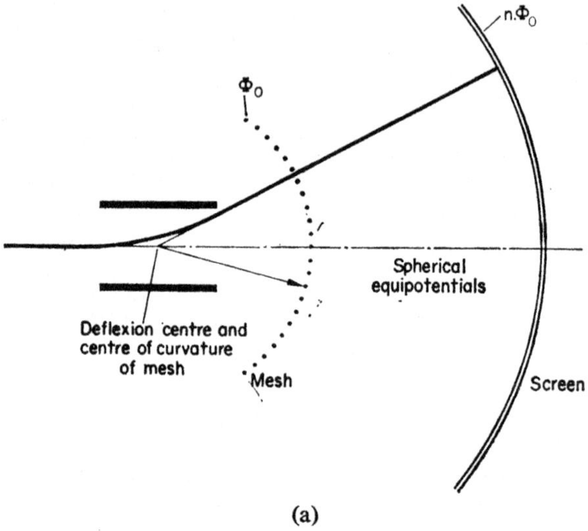

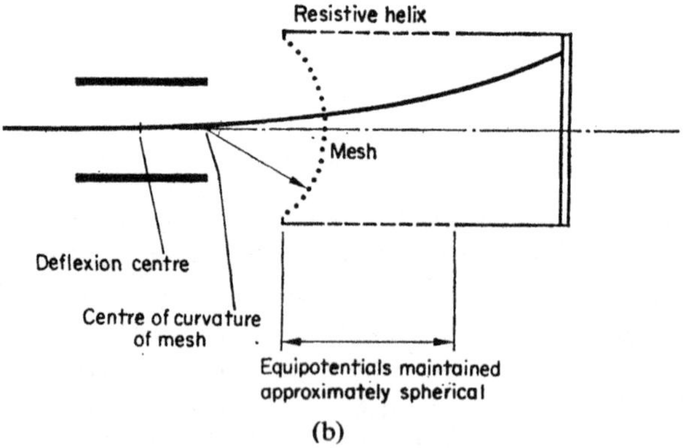

FIG. 38. Post-acceleration deflexion using mesh lens. (a) Centre of curvature of mesh coincides with deflexion centre; equipotentials between mesh and screen are spherical. Deflexion independent of n. (b) Centre of curvature of mesh beyond deflexion centre; equipotentials approximately spherical for most of acceleration. Deflexion increases with n.

for focusing electron beams suggested, but until recently not exploited. The success of this arrangement owes a great deal to modern techniques for making transparent grids, with well-defined cross-sectional forms designed to reduce secondary emission towards the screen. Secondary electrons on the gun side are removed by the use of a suitable bias potential. It has also been found that the flexibility in electron optical design which springs from the use of a mesh lens can be exploited to reduce the deflexion defocusing effects which are the subject of the next section.

7.4. Deflexion aberrations

In an electron beam device such as the c.r.t., the electron optical system will produce a spot on the target plane, which is generally designed to be as small as possible. Its size will be determined by some combination of thermal velocity effects and the aberrations of the focusing lens. An ideal deflecting system would then produce a deflexion of this spot by an amount strictly proportional to the magnitude of the deflecting voltage (or current), and moreover would lead to no change in the size or shape of the spot. Practical deflectors introduce aberrations which lead to departure from both these ideals.

The nature of these departures is very similar to the errors of lenses discussed in section 4. For example, a departure from linearity in the relationship between deflecting voltage and the deflexion is directly comparable with distortion, and effects corresponding to curvature of field coma and astigmatism are also readily observed. It is therefore entirely possible to analyse these defects in much the same manner as in the case of lenses. Such a course is not, however, much to be recommended for two reasons: firstly, the deflecting fields lack circular symmetry; although they do possess mirror symmetries, the result is a great increase in the complexity of the aberration analysis. Secondly, one is in practice frequently concerned with very large deflexions (e.g. in excess of 60° in some picture tubes). Under these conditions terms of higher order than those considered in third order aberration theory may dominate the situation. Their inclusion makes the frontal attack even more cumbersome. If very detailed knowledge of the aberrations is required, direct computer trajectory determination will usually be the best approach. In many cases one is, however, primarily concerned with the order of magnitude of the defects, which will suffice for deciding on the nature of the correction devices which need to be incorporated.

It is none the less instructive to examine some simple and highly idealized situations—even though they will hardly lead to usable quantitative information. We will accordingly again turn to the simple parallel plate electrostatic deflector, with the fringing field neglected. It is necessary to treat such an idealized case with some caution. For example, taken literally, one would conclude that the

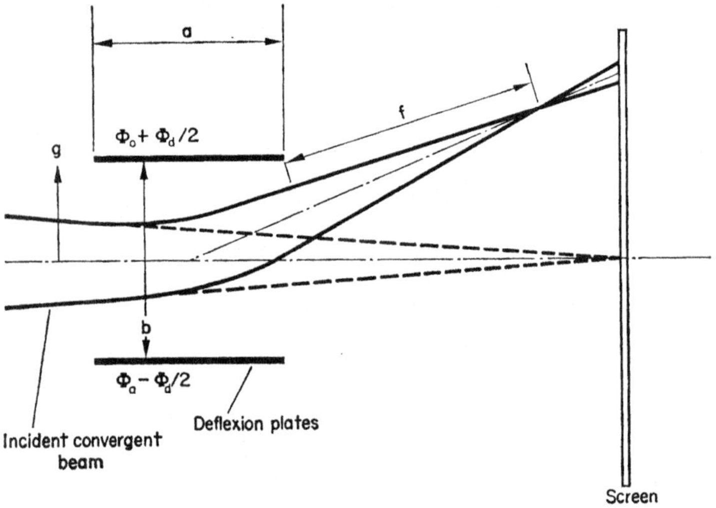

FIG. 39. Deflexion defocusing in deflexion plane.

deflected electron emerges with more energy that it had on entry, an embarrassment which we must suffer as a penalty for neglecting the fringing fields. For this reason this model cannot predict deflexion non-linearity. However, it can give us some insight into the astigmatism introduced by the deflector (Fig. 39). Suppose a parallel beam of radius r_0 enters the deflecting region. As we found in section 7.1, the deflexion angle θ produced is inversely proportional to the potential at the point of entry of the electron considered. If we regard the deflecting fields as weak, we can confine our attention to this deflexion angle. From (7.4) we can then write

$$\theta \simeq \frac{\Phi_d}{\Phi_0 - (y/b)\,\Phi_d} \cdot \frac{a}{b}.$$

This dependence of θ on y implies the focusing action which we are seeking. The parallel rays will be brought to a focus in a distance f, where

$$\frac{1}{f}=\frac{d\theta}{dy}=\frac{b^2}{a}\cdot\left(\frac{\Phi_d}{\Phi_0}\right)^2.$$

Thus the strength of the focusing effect is proportional to the square of Φ_d, and hence to the square of the final deflexion. The convergence arises in one plane only, so that the deflector acts as a pure cylindrical lens. This effect will reveal itself in a spot which is astigmatic when deflected. If the main focusing lens strength is varied, focus will be achieved in the plane at right-angles to that of the deflectors, before that in the plane of the deflectors.

A similar effect does not arise in the case of the idealized uniform magnetic field deflector,† although fringing fields will undoubtedly generate astigmatic effects.

Practical deflecting fields are, of course, much more complex than those considered above—particularly in the case of magnetic deflexion. The knowledge of the fields which must precede trajectory computation can be obtained, as discussed in section 3.2, either by experimental measurement, or by computer solution of the field problem. The choice is largely one of economics. However, comparing the case of lenses with that of beam deflexion devices such as the c.r.t., the region to be mapped in the latter case is typically much larger, and the required accuracy somewhat less. The scales are therefore weighted somewhat more towards experimental measurement in this case. A detailed study of a c.r.t. deflexion system has recently been reported(104). The field was measured in two planes using a Hall probe and the field at other planes computed by vector addition. Trajectories of electrons initially on the axis were then computed and led to detailed conclusions on deflexion non-linearity. In order to compute other deflexion aberrations one would, in addition, have to compute trajectories not initially on the axis. To predict the astigmatism it would suffice to confine the computations to rays whose axial displacement was at right-angles to the direction of the magnetic field on the axis. In this case the trajectories remain in one axial

† It is a well-known property of the uniform magnetic field that it will bring a parallel beam to a focus. However, this property depends on a difference in the transit time between various electrons in the beam to the focal point. In the case of the idealized magnetic deflector the transit time through the field is the same for all electrons, so that no focusing action arises.

plane, and little extra complexity is involved. In order to obtain a complete picture of the aberrations, one would have to work with rays with an initial displacement in an arbitrary direction, and the trajectories would not then lie in a single plane. Such computations would then be substantially longer, but computing costs are still likely to be sufficiently low to encourage the use of such direct methods.

The main aberrations which the designer will seek to minimize or compensate are, then, deflexion non-linearity, deflexion defocusing and astigmatism. The simplest method of reducing deflexion non-linearity is by the use of auxiliary static quadrupole magnets[105].

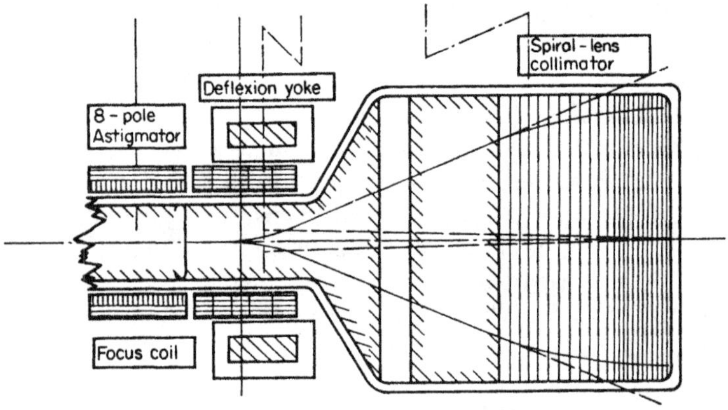

FIG. 40. Correction of focus and astigmatism in high resolution tube (after Schlesinger and Wagner, ref. 106).

These can be designed to compensate the distortion (usually "pin cushion") due to the action of the main deflecting coils, and are frequently used on picture tubes. The reason why the compensation can be static is, of course, that the compensating fields come into play only when the electron beam enters the region where they are strong —i.e. when they are strongly deflected. Indeed, the technique can be used for increasing the deflexion ("scan magnification"). It is naturally not possible to effect any similar static correction for astigmatism or deflexion defocusing, and for many applications it is in any event not necessary. However, in critical applications such as high definition microspot tubes, the errors must be compensated by means of special correcting fields energized by the same current (or voltage) which is controlling the main deflecting field. An example of such an application which was recently reported[106] required the maintenance of an 8 μ spot size over a screen diameter of 10 cm. Without compensation, deflexion defocusing increases the spot size 160 μ at extreme deflexions. Compensation of this error then still leaves a spot size of around 60 μ attributable to astigmatism. The method of compensation adopted in this study which led to negligible degradation of the deflected spot size is shown in Fig. 40. The focus is corrected by means of the auxiliary focus cell. The astigmatism is corrected by means of an auxiliary quadrupole field, which in accordance with an earlier suggestion by Hutter and Harrison[107] is located just prior to the main deflexion coils. Although the required correction field is quadrupolar, the correct orientation depends on the direction of deflexion of the beam. The correction device is therefore made octupolar, so that by varying the relative current to adjacent poles the orientation of the quadrupole field can be controlled. It turns out that the correction currents needed for both astigmatism and focus correction are quadratic functions of the currents in the main deflexion coils. In the work reported, these were obtained from a special analogue computer. They could, of course, be computed digitally, and one might envisage that a design realized in microcircuits could be sufficiently compact and inexpensive to justify large-scale application.

REFERENCES

1. STURROCK, P., *Static and Dynamic Electron Optics.*
2. BUSCH, H., Berechnung der Bahn von Kathoden Strahlen in axial symmetrischen elektromagnetischen Feld, *Ann. Phys.* **81**, 974–93 (1926).
3. BUSCH, H., Uber die Wirkungsweise der Konzentrierungsspulebe dier Braunschen Rohre, *Arch Elektrotech.* **18**, 583–94 (1927).
4. GABOR, D., Die Entwicklungsgeschichte des Elektronenmikroskops, *Elektrotech. Z.* **78**, No. 15, 522–30 (1957).
5. DUPOUY, G., FABRE, R. and PERRIER, F., *Compt. Rend. Acad. Sci.* **5**, 252 (1961).
6. DE BROGLIE, M. L., *Optique Electronique et Corpusculaire*, Hermann, Paris, 1950.
7. See, for example, DE BROGLIE, M. L., *Optique Electronique et Corpusculaire*, p. 101, Hermann, Paris, 1950.
8. PICHT, J., *Annls Phys.* **15**, 926 (1932).
9. STURROCK, P., *Static and Dynamic Electron Optics.*
10. PIERCE, J. R., *Theory and Design of Electron Beams*, p. **100**, Van Nostrand 1954.
11. GLASER, W., *Z. Physik*, **117**, 285, 315 (1941); **118**, 264 (1941).
12. BERNARD, M. Y. and GRIVET, P., *Annls Radioélect.* **6**, 3 (1952).
13. GRIVET, P., *Compt. Rend. Acad. Sci.* **233**, 921 (1951).
14. LENZ, F., *Ann. Phys.* **9**, 245 (1951).
15. LENZ, F., *Optik*, **7**, 243 (1950).
16. STURROCK, P., *Compt. Rend. Acad. Sci.* **233**, 401 (1951).
17. GLASER, W. and SCHISKE, P., *Optik*, **11**, No. 9, 422–43 (1954); **12**, No. 5, 233–44 (1955).
18. ARCHARD, G. D., *Brit. J. Appl. Phys.* **7**, 330–2 (1956).
19. KLEMPERER, O. and WRIGHT, W. D., *Proc. Phys. Soc.* **51**, 296 (1939).
20. SPANGENBERG, K. and FIELD, L. M., *Proc. I.R.E.* **3**, 138–44 (1942).
21. DOSSE, J., *Z. Physik*, **117**, 722–53 (1941).
22. HEISE, F. and RANG, O., *Optik*, **5**, 201 (1949).
23. FERT, C. and GAUTIER, P., *Compt. Rend. Acad. Sci.* **233**, 148 (1951).
24. GAUTIER, P., *Compt. Rend. Acad. Sci.* **242**, No. 13, 1707–10 (1956).
25. PEARSON, A., *J. Sci. Instr.* **39**, 8–10 (1962).
26. ADAMS, W. G., *Proc. Roy. Soc., London.* **23**, 280–84 (1875).
27. LIEBMANN, G., *Adv. Electronics*, **2**, 102–46 (1950).
28. SANDER, K. F., OATLEY, C. W. and YATES, J. G., *J.I.E.E.*, Pt. III, **99**, 169 (1952).
29. PIZER, H. I., YATES, J. G. and SANDER, K. F., *J. Electr.* **2**, 65–86 (1956).
30. BREWER, G. R, Formation of high density electron beams, *J. Appl. Phys.* **28**, 7 (1957).
31. LIEBMANN, G., *Brit. J. Appl. Phys.* **1**, 92 (1951).
32. HECHTEL, J. and SEEGER, A., *Proc. I.R.E.* **49**, 933–40 (1961).
33. SOUTHWELL, R. V., *Relaxation Methods in Engineering Science*, Clarendon Press, Oxford, 1940.
34. MOTZ, H. and KLANFER, L., *Proc. Phys. Soc.* **58**, 30 (1946).
35. KORMES, M., *Rev. Sci. Instr.* **14**, 248–50 (1943).
36. KIRSTEIN, P., KINO, G. S. and WATERS, W. E., *Space Charge Flow*, McGraw-Hill, (1967).
37. GABOR, D., *Nature*, **139**, 373 (1937).

38. LANGMUIR, D. B., *Nature*, **139**, 1067 (1937).
39. SMYTHE, W. R., RUMBAUGH, L. H. and WEST, S. S., *Phys. Rev.* **45**, 724 (1934).
40. WALKER, G. B., *Proc. I.E.E.* **96**, 319 (1949).
41. ZWORYKIN, V. K. and RAJCHMAN, J. A., *Proc. I.R.E.* **27**, 558 (1939).
42. LIEBMANN, G., *Proc. Phys. Soc.* B, **62**, 753–72 (1949).
43. BURFOOT, J. C., *Brit. J. Appl. Phys.* **3**, 22–4 (1952).
44. HAINES, M. E. and VINE, J., *Proc. I.E.E.*, **106** B, 517–27 (1959).
45. CARRE, B. A. and WREATHALL, W. M., *Proc. Brit. I.R.E.* **27**, No. 6, 446–54 (1964).
46. CARRE, B. A., *Computer J.* **4**, 73–80 (1961).
47. ZWORYKIN, V. K., MORTON, G. A., RAMBERG, E. G., HILLIER, J., and VANCE, A.W., *Electron Optics and the Electron Microscope*, p. 450, Wiley, New York, 1945.
48. OLLENDORF, F., *Elektronik des Einzelelektrons*, p. 340, Springer, 1955.
49. SPANGENBERG, K., *Vacuum Tubes*, pp. 370 *et seq.*, McGraw-Hill, 1948.
50. RAMBERG, E. G., *J. Appl. Phys.* **13**, 582 (1941).
51. REGENSTREIF, E., *Annls Radioélect.* **6**, 24 (1951).
52. LIEBMANN, G. and GRAD, E. M., *Proc. Phys. Soc.* **64** B, 956–71 (1951); LIEBMANN, G., *Proc. Phys. Soc.* **66** E, 448–58 (1953); **65**, 94–108 (1952); **68** B, 679–81 (1955); **68** B, 682–5 (1955); **68** B, 737–45 (1955).
53. DE BROGLIE, M. L., *Optique Electronique et Corpusculaire*, pp. 131–2, Hermann, Paris, 1950.
54. SCHERZER, O., *Z. Physik*, **101**, 593–603 (1936).
55. TRETTNER, W., *Optik*, **16**, 155–84 (1959).
56. See, for example, COSSLETT, V. E., *Electron Optics*, Oxford, 1946.
57. See, for example, GRIVET, P., *Optique Electronique*, p. 98, Bordas, 1955.
58. GLASER, W., *Grundlagen der Elektronen Optik*, Springer, 1952.
59. WEITZCH, *Optik*, **15**, No. 8, 492–9 (1958).
60. GABOR, D., *Electr. Eng.* **15**, 295 (1942); 328 (1943); 372 (1943).
61. HAUFE, G., *Optik*, **15**, No. 9, 521–37 (1958).
62. ASH, E. A., *J. Appl. Phys.* **26**, 327 (1955).
63. SCHERZER, O., *Optik*, **2**, 114 (1947); **5**, 497 (1949).
64. HILLIER, J. and RAMBERG, E. G., *J. Appl. Phys.* **18**, 48 (1947).
65. SEELIGER, R., *Optik*, **5**, 490 (1949); **8**, 311 (1951).
66. BURFOOT, J. C., *Proc. Phys. Soc.* B, **6**, 775–92 (1953); WHITMER, R. F., *J. Appl. Phys.* **27**, 808 (1956); ARCHARD, G. D., *Proc. Phys. Soc.* B, **72**, 135–7 (1958); SEPTIER, A. and VAN ACKER, I., *Compt. Rend. Acad Sci.* **251**, 346 (1960).
67. GABOR, D., *Proc. Phys. Soc.* B, **64**, 244 (1951).
68. JULL, G. W., *Passage of Medium Energy Electronics through solids. Ph.D. Thesis*, University of London (1955).
69. KNOLL, M. and RUSKA, E., *Annls Phys.*, series 5, **12**, 607–61 (1932).
70. LAW, H. B., DAVNE, L. and RAMBERG, E. G., *R.C.A. Rev.* 603–22 (1961).
71. BERTEIN, F., *Annls Radioélectr.* **3**, 49 (1948); HAINES, M. E. and MULVEY, T., *J. Sci. Instr.* **31**, 326 (1954).
72. STOJANOW, P. A., *Proceedings of 4th International Electron Microscopy Conference*, pp. 61–6, Springer, 1960.
73. MÖLLENSTEDT, G., *Optik*, **5**, 499–517 (1949).
74. HAINES, M. E., *The Electron Microscope*, Chap. VI, Spon, 1961.
75. MAYER, L., Electron mirror microscopy of magnetic domains, *J. Appl. Phys.* **28**, No. 9, 975, 983 (Sept. 1957).
76. MAYER, L., On electron mirror microscopy, *J. Appl. Phys.* **26**, 1228 (1955).
77. BARNETT, M. E. and NIXON, W. C., Topographical mirror microscopy,

Proceedings of 4th International Electron Microscopy Conference, p. 37, Prague, 1964.
78. OATLEY, C. W., NIXON, W. C. and PEASE, R. F. W., Scanning electron microscopy, *Adv. Electronics Electron Phys.* **21**, 181 (1965).
79. MUELLER, Z. W., Field ionization and field ion microscopy, *Adv. Electronics*, **7**, 83–179 (1960).
80. ZWORYKIN *et al.*, *Electron Optics and the Electron Microscope*, Section 13.9, Wiley, 1945.
81. VINE, J., Electron optical ray training in cathode lenses, *Trans. I.E.E.E.* ED–13, No. 7, 544–51 (July 1966).
82. ASH, E. A., Limiting current densities in the presence of a magnetic field, *J. Appl. Phys.* **35**, No. 2, 298–301 (Feb. 1964).
83. WILCOCK, W. L., EMBERSON, R. L. and WEEKLY, B., *Trans. I.R.E.* NS–7 No. L–3, 126 (1960).
84. MCGEE, MCMULLAN, D. and KAHAN, E. (ed.), Photoelectric image devices, *Adv. Electronics Electron Phys.* **22**, (1966).
85. OATLEY, C. W., NIXON, W. C. and PEASE, R. F. W., Scanning electron microscopy, *Adv. Electronics Electron Phys.* **21**, 181 (1965).
86. COSSLETT, V. E. and NIXON, W. C., *X-ray Microscopy*, Cambridge U.P., 1960.
87. See, for example, *Proceedings of Symposium on Electron Beam Techniques for Microelectronics in Microelectronics and Reliability*, Vol. 4, No. 1, pp. 1–122, March 1965.
88. CUTLER, C. L. and HAINES, M. E., Thermal velocity effects in electron guns, *Proc. I.R.E.* **43**, 307–14 (1955).
89. LANGMUIR, D. B., Theoretical limitations of cathode ray tubes, *Proc. I.R.E.* **25**, 977 (Aug. 1937).
90. GABOR, D., presented in *Electron Optics*, by L. Myers, London, 1939.
91. PIERCE, J. R., *J. Appl. Phys.* **10**, 715 (1939).
92. PIERCE, J. R., *Theory and Design of Electron Beams* (2nd ed.) Chap. VIII, Van Nostrand, 1954.
93. ASH, E. A., Limiting current densities in the presence of a magnetic field, *J. Appl. Phys.* **35**, No. 2, 298–301 (Feb. 1964).
94. ALFVEN, H., *Cosmical Electrodynamics*, Oxford, 1950.
95. HUTTER, R. G. E., Electron beam deflection, *J. Appl. Phys.* **18**, 740–8, 797–810, Aug./Sept. 1947.
96. GLASER, W., *Grundlagen der Elektronenoptik*, Chap. XXIII, Springer, 1952.
97. KLEMPERER, O., *Electron Optics*, p. 330, Cambridge, 1953.
98. SCHWARTZ, E., Zum Stande des Nachbeschleunigungsproblem by Kathodenstrahlroehren, *Fernsch A.G.* **1**, 19 (1938).
99. GUNDERT, E. and SCHAFFERNICHT, W., New cathode ray tubes of higher brightness and deflection sensitivity, *Elektrophys. Rdsch.* **11**, No. 4, 107–8 (April 1957).
100. PIERCE, J. R., *Theory and Design of Electron Beams*, Section 8.7, Van Nostrand, 1954.
101. LAW, H. B., DAVNE, L. and RAMBERG, E. G., The enhanced scan, post-acceleration kinescope, *R.C.A. Rev.*, **22**, 603–22 (Dec. 1961).
102. SIEGEL, F. G., A new 50 mc oscilloscope based on an advanced CRT design, *Hewlett Packard Journal*, **13**, No. 8, 1–7 (April 1962).
103. KRUGER, W. (private communication).
104. BLOOMSBURGH, R. A. and KING, J., Computer determination of electron paths in magnetic deflection fields, *Proceedings of 6th Symposium on Electron Beam Technology*, pp. 14–31, Boston, 1964.

105. JOHNSON, K. E., Picture tubes for television displays using quadruple scan magnification, *Radio and Electronic Engineering* (*G.B.*), **28**, 115–28 (Aug. 1964).
106. SCHLESINGER, K. and WAGNER, R. A., Correction of deflection aberration by analog computer, *Trans. I.E.E.E.*, ED–**12**, No. 8, 478–84 (Aug. 1965).
107. HUTTER, R. G. E. and HARRISON, S. W., A method of reducing deflection defocusing in cathode ray tubes, *J. Appl. Phys.* **21**, 84–9 (Feb. 1950).
108. BARNETT, M. E. and NIXON, W. C., A mirror electron microscope using magnetic lenses, *J. Sci. Instr.* **44**, 893–8 (Nov, 1967).
109. BULLOUGH, R., HEATH, A. C. and VINE, J., A general purpose digital computer programme for the evaluation of electron lens parameters. *A.E.I. Engineering*, pp. 42–6 (Dec. 1963).

PART 5

HEAVY CURRENT ELECTRON BEAMS

A. H. BECK

Engineering Laboratory, Cambridge

PART 5
HEAVY CURRENT ELECTRON BEAMS

A. H. BECK
Engineering Laboratory, Cambridge

MODERN electron beam devices utilize an extremely wide range of currents. At the high voltage end of the scale we encounter on the one hand electron microscopes working at well over 100 kV with beam currents of 50–100 μA and on the other hand klystrons designed for high powered pulse operation in the same voltage range in which the beam current is 100 A. As an example of an extremely low voltage electron beam device we have the Adler type of parametric amplifier in which the beam current is a few tens of microamperes at voltages of the order of 10 V. In the tubes, klystron and parametric amplifier the current must be made to travel in the near vicinity of a high frequency circuit for most of the length of the beam. This is because high frequency fields diminish very rapidly as the distance between the point of measurement and the metallic walls of the circuit increases. Therefore, the ideal beams for high frequency valves are long cylinders or long beams of rectangular cross-section. By contrast, the microscope beam is convergent/divergent so that it is only focused to the axis of the device over a short portion of the total axial length.

In these circumstances it is easy to see that the influence of space-charge forces will be very much less in the low current convergent/divergent beams of conventional electron optics than in the high current quasi-cylindrical beams of modern tube technique. We are thus led to define a new science of heavy current electron optics dealing with beams in which the forces due to space-charge repulsion are major factors in determining the trajectories of the individual electrons forming the beam. Logically, we ought to begin the study of this subject by establishing criteria which should be applied to allow us to know whether we are in fact dealing with a situation in which space-charge forces are dominant. The precise establishment of such criteria usually require that the geometry of the system under consideration be known, but Pierce[1] has established the following brightness relation which gives the maximum current density J in a spot formed by a system of spherical lenses in terms of the cathode density J_0, the potential of the region in which the spot is formed and the semi-angle of convergence to the spot θ. This relation is

$$J = J_0 E_i \left(1 + \frac{eV}{kT}\right) \sin^2 \theta$$

$$\doteqdot J_0 E_i . \frac{11{,}606}{T} \sin^2 \theta. \qquad (1)$$

Here, E_i is an efficiency factor which depends on the way in which the spot diameter is defined. In particular, if the spot is defined as containing 90 per cent of the total beam current, $E_i = 39$ per cent. Equation (6.7), as shown on p. 485, depends on the Maxwellian distribution of initial electron velocities and is therefore a purely thermal limitation.

Let us now put $E_i = 0.39$ and take $T = 1160°\text{K}$, which is not unreasonable for tubes using beams, then eq. (1) reduces to

$$J = 3.9\ V \sin^2 \theta . J_0. \qquad (2)$$

For the sake of concreteness, take $V = 1$ kV and $\sin \theta = 0.1$. Then $J = 39\, J_0$ and we can conclude that if the maximum current density at any point in our device is small compared with $39\, J_0$, then the beam will be mainly controlled by space charge. When the current density anywhere approaches the limiting value, thermal effects dominate the effects of space charge. For cylindrical lens systems (line focus) the corresponding result is

$$J = J_0 \left\{ 2\left(\frac{eV}{\pi kT}\right)^{\frac{1}{2}} + \exp\left(\frac{eV}{kT}\right)\left[1 - \text{er} + \left(\frac{eV}{kT}\right)^{\frac{1}{2}}\right] \sin \alpha \right\}$$

$$\text{where er} + x = \frac{2}{\sqrt{\pi}} \int_0^x \exp(-u^2)\, du.$$

The question of limiting current densities will become much clearer when we have described the behaviour of some typical focusing systems.

A heavy current electron optical device can conveniently be divided into two major regions: the electron "gun", i.e. the cathode and acceleration region in which the electrons are accelerated to their final potential, and the beam region, in which, as we have said, a quasi-cylindrical beam must be maintained or, as it is frequently but loosely stated, "focused". Electron optical devices can be classified into three types, depending on whether the beam is maintained by purely electrostatic means, by purely magnetic means, or by a combination of both electric and magnetic fields. Electric fields must, of course, be present in the gun region to accelerate the electrons away from the cathode surface. Examples of the three types of beam maintenance are:

1. Electrostatic. The use of a large number of apertured discs at alternative positive and negative potentials.
2. Magnetic. The provision of a strong axial magnetic field by the use of a solenoid or suitable permanent magnet.
3. Combined. One example was described by Chang[(2)]. The beam is set spinning by allowing it to emerge from an axial magnetic field region into a region free from magnetic field. The spinning beam is maintained by electrostatic lenses as described in the first example, but the stability of the flow is much improved.

In addition "space-charge focused" devices are also freely used. Here, the device is short enough for an adequate current to be passed without any special form of beam maintenance. Instead, the beam is made to converge at exit from the gun. It then converges to a "throat" of minimum diameter where all the electrons move parallel to the axis and thence diverges under the influence of space-charge forces. The divergence can be calculated from a beam spread curve. All these processes occur in a region where there are no d.c. fields.

1. SCALING

One of the most useful concepts in the development of electron optical systems is that of scaling, which allows us both to test electron guns at voltages and currents different from those at which eventual operation is desired and to test guns of different dimensions from those finally required.

The test voltage is usually lower than the operating voltage so that power inputs to the electrodes are reduced and troubles due to gas production, insufficient cathode emission and the like can be avoided.

Voltage scaling is accomplished using the generalized space-charge limited current expression

$$I = \frac{kV^{3/2}}{S^2}, \tag{3}$$

where I = beam current,
k = constant,
V = applied voltage,
S = a characteristic length.

Usually, we may use the simplified form of (3), viz. $I = pV^{3/2}$.

It will be realized from our opening remarks on the nature of current limitations in electron beams that as the test voltage is lowered, a value will eventually be reached at which effects due to thermal velocity spread dominate those of space charge. The mean thermal energy for electrons originating from oxide or dispenser cathodes is less than 0.2 eV. The corresponding velocity is therefore very small compared with the ordered velocity of a 300 V beam and we may conclude that voltages of less than 300 V should not be used to predict gun performance at much higher voltages. In the case of very convergent guns much higher minimum test voltages may be required.

2. DIMENSIONAL SCALING

Another form of scaling is equally important. (a) If all the electrode dimensions are multiplied by a given scale factor all the electron trajectories are multiplied by the same scale factor. A corollary is (b) an electron trajectory remains unchanged if the voltages on all the electrodes are changed by the same factor. (This is more general than the straightforward voltage scaling as it applies to any number of electrodes.)

The importance of dimensional scaling arises particularly in microwave valve development. Suppose that an efficient gun/beam system has been designed and operates in a valve at wavelength λ_1. The requirement then arises for a valve to work at wavelength λ_2. If all the dimensions of the gun/beam system are multiplied by the ratio λ_2/λ_1 (normally <1) the second valve will have the same current and beam efficiency as the first. Two qualifications have to be made. The first concerns the cathode current density. In the second valve the cathode area is reduced by the factor $(\lambda_2/\lambda_1)^2$ but the current remains the same. The current density is therefore increased by $(\lambda_1/\lambda_2)^2$ and this ratio must therefore be limited so that the final density does not exceed the value allowed by the cathode properties. The second qualification is again concerned with the thermal limitation. If the final beam density is very high, the beam may be required to operate at a voltage characteristic of thermal limitation and the resulting beam formation will not therefore be similar to the original space-charge limited beam. However, in many cases dimensional scaling is allowable and is obviously a procedure which saves much development time. It is, perhaps, not superfluous to comment that dimensional scaling must be applied to *all* the dimensions of the scaled gun, including, for example, the thickness of thin edges, etc., if these take part in the field formation.

3. ELECTRON GUNS

Historical. Early heavy current electron guns, designed before about 1940, were enlarged versions of electron guns used in cathode ray tubes. They were often elaborate, gave relatively small currents and were badly aberrated. They were often designed using field plots made in an electrolytic trough† to compute, by graphical means, the trajectories of typical electrons. Sometimes, various approximation techniques, for example the relaxation technique of Southwell, were used to correct the computed trajectories for space charge. Few criteria existed by which the goodness of a gun could be assessed. However, in spite of these difficulties some guns of fairly good performance were designed and some useful general principles were discovered. A typical gun of this era is shown in Fig. 1. It was used in an early klystron amplifier.

This situation was transformed by a revolutionary idea due to Pierce[(3)] who, instead of working from a given structure and inquiring what trajectories were produced, inquired into the structures needed to produce desirable trajectories, for example, the trajectories on the surface of a truncated cone needed in the converging spherical electron gun which is almost universal today. Many years earlier Langmuir and Blodgett[(4)] had calculated the potential distributions *inside* space-charge limited flows in planar, cylindrical and spherical geometries. The problem was, therefore, to determine electrode

† Now, a digital computer would be used.

structures which would continue these potentials outside the beam, giving the correct potential variation along the boundary between the electron beam and the remaining electron-free space.

In some cases this could be done analytically, and much time has since been spent, not very usefully, in extensions of this part of Pierce's work. In other, more important, cases experiment in the electrolytic trough (or other computer) was necessary. At present we may confine ourselves to the practical achievement, which showed

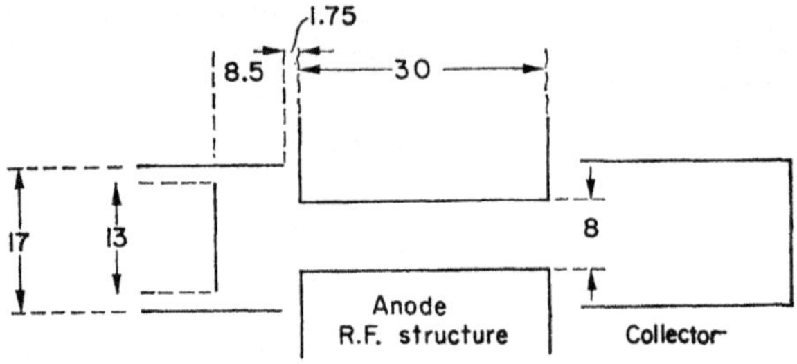

FIG. 1. The focus cylinder ran at cathode potential. The discrepancy between the perveance figures is partly due to cathode emission and partly due to oscilloscope calibrations which were more reliable at the higher currents.

$$\left.\begin{aligned} V_a &= 12 \text{ kV} \\ I_{coll} &= 1.12 \text{ A} \\ I_a &= 0.06 \text{ A} \\ \eta_g &= 96 \text{ per cent} \end{aligned}\right\} \quad p \doteqdot 0.9 \times 10^{-6}$$

$$\left.\begin{aligned} V_a &= 4 \text{ kV} \\ I_{coll} &= 290 \text{ mA} \\ I_a &= 9 \text{ mA} \\ \eta_g &= 97 \text{ per cent} \end{aligned}\right\} \quad p \doteqdot 1.2 \times 10^{-6}$$

that guns of the general form shown in Fig. 2 could be made and would give perveances† of up to 0.5×10^{-6} and that well over 90 per cent of the current would pass through the anode aperture. To increase the current at a given voltage, i.e. to increase the perveance, one must either increase the cathode area, involving an increase in the maximum angle of convergence, or decrease the anode–cathode spacing, or combine both these modifications.

It was soon found that it was impossible to design efficient guns with perveances of more than about 10^{-6} using the Pierce technique

† Most modern electron guns behave as simple diodes giving a current–voltage characteristic $I = pV^{3/2}$. The constant p in this relation is the perveance, often expressed in micropervs. A gun with a perveance of 1 microperv gives 1 mA at 100 V and 1 A at 10 kV.

and this in turn was due to the fact that the necessary aperture in the anode perturbed the field inside the beam to such an extent that design methods based on the assumption of spherical flow were no

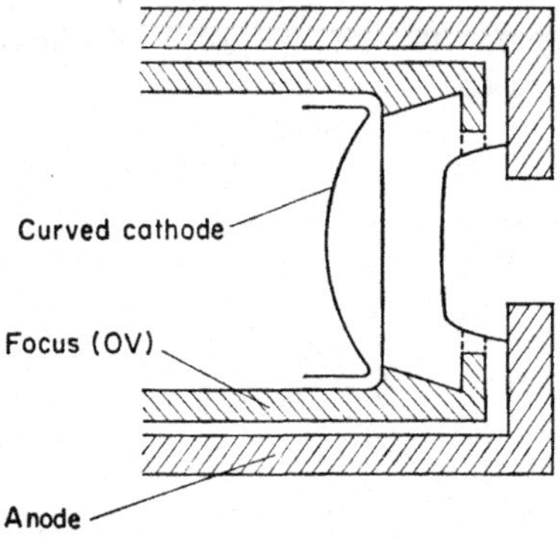

FIG. 2. Pierce point focus gun, as developed by Samuel, $p \doteqdot 1.6 \times 10^{-6}$, $\eta_g = 96$ per cent. Calculated $p = 6 \times 10^{-6}$.

longer correct. However, the behaviour of the electrons in the immediate vicinity of the cathode is not affected and using the tank for space-charge corrected trajectory plotting it was possible to modify the anode region without altering the cathode geometry.

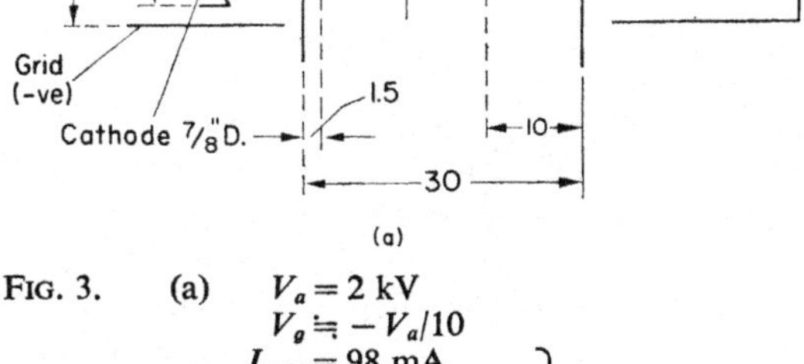

FIG. 3. (a) $V_a = 2$ kV
$V_g \doteqdot -V_a/10$
$I_{coll} = 98$ mA
$I_a = 2$ mA
$\eta_g = 98$ per cent
} $p = 1.1 \times 10^{-6}$

Semi-empirical gun designs of this type were used during World War II with perveances up to 1.7×10^{-6}. Figure 3 shows examples described by Beck, Jackson and Preddy[(5)] in an unpublished report which was, however, widely circulated. Figure 3a shows a gun which appears to have been scaled to much larger and to much smaller sizes.

At present, solid beam guns of this type with perveances in excess of 3×10^{-6} are not available. For this reason, among others, there is now a good deal of interest in guns producing annular beams which may be expected to give perveances higher by a factor of ten.

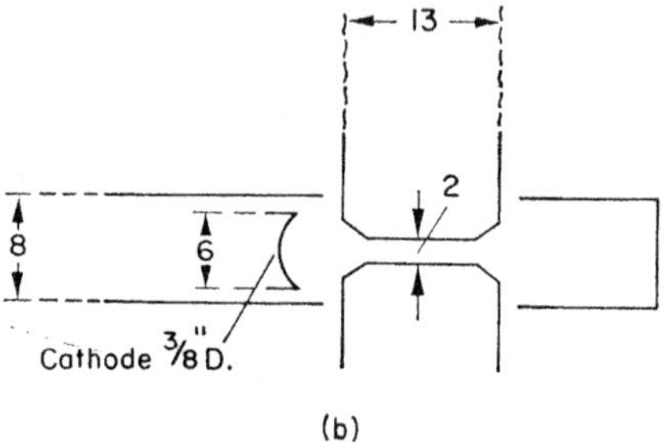

FIG. 3. (b) $V_a = 2$ kV, $V_g = -V_a/5$, $I_{coll} = 98$ mA, $I_a = 6$ mA, $\eta_g = 94$ per cent; $p = 1.1 \times 10^{-6}$

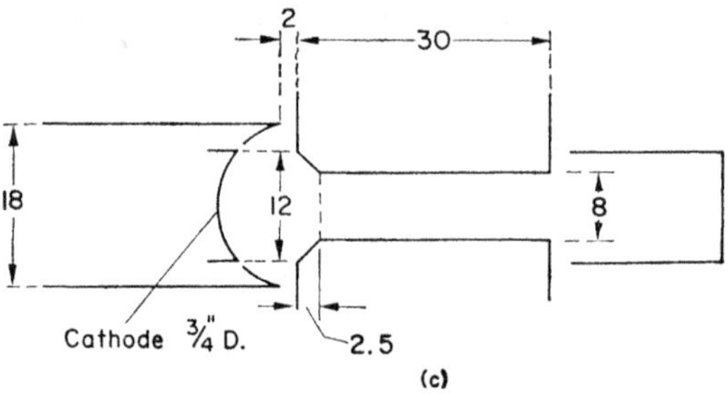

FIG. 3. (c) $V_a = 12$ kV, $I_{coll} = 2$ A, $I_a = 0.18$ A, $\eta_g = 92$ per cent; $p = 1.7 \times 10^{-6}$

Figure 4 shows a gun designed by the author which has a perveance of nearly 3×10^{-6}. The illustration includes electron trajectories, plotted by Barber and Sander[(6)] using the Cambridge automatic electrolytic tank, which has facilities for the injection of currents into the model to simulate the effects of space charge. Trajectories with and without space charge are shown and these illustrate the way in which space-charge forces have been used to produce a "throat" in

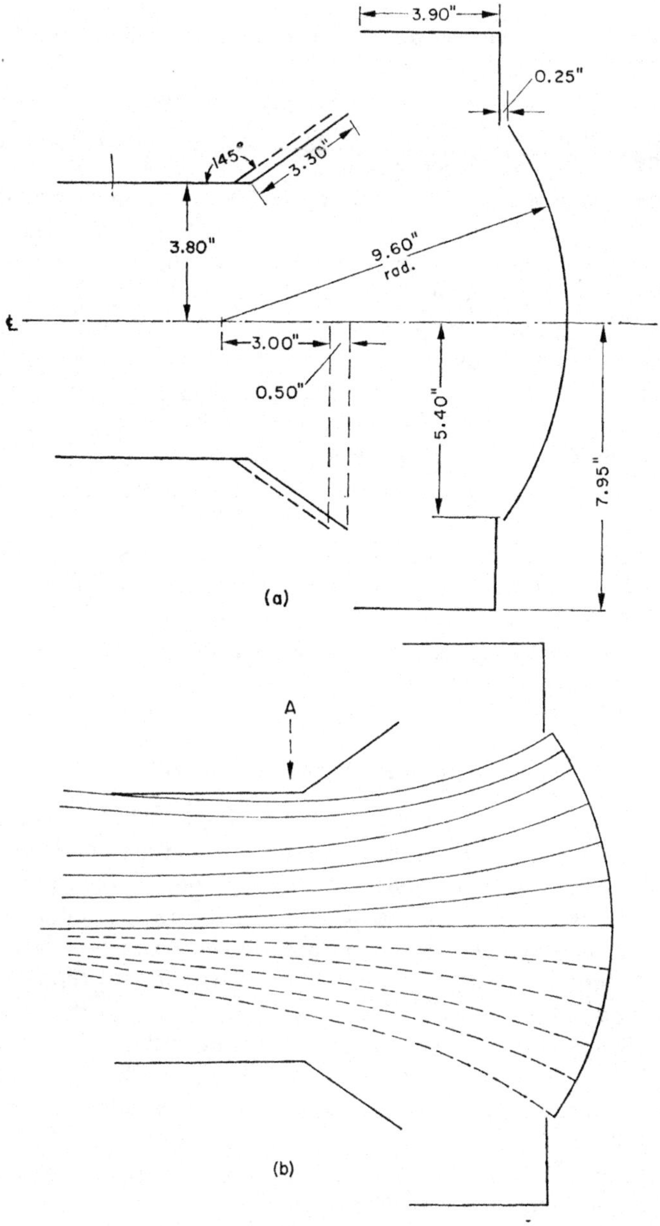

FIG. 4. (a) Gun dimensions. Two positions of the anode are shown. The calculated perveance in the nearer was 2.9×10^{-6}, in the further 2.4×10^{-6}. (b) Gun trajectories. Without space charge - - - - -. With space charge ———.

which the electrons closely approximate to laminar flow. We shall expand these remarks later.

It should be noted that many other properties besides perveance are important in practical electron guns. One such property is the area convergence ratio which specifies the ratio of cathode area to the (assumed cylindrical) beam area. Since the maximum available cathode current density is known, from the properties of the particular emitter surface used and the operating temperature, the convergence ratio gives the maximum available beam current density. This latter quantity is one of the major parameters of the valve design.

It is found, empirically, that it is easier to obtain high perveances from guns with a low convergence than it is from high convergence guns. In fact, Beck and Deering[(7)] have described a diode gun, in which the convergence is substantially unity, which has a perveance of 6×10^{-6}. This gun, developed for a pulsed klystron, uses a screen electrode pulsed to one-third of the main beam voltage to control the beam current. This screen only intercepts about $\frac{3}{4}$ per cent of the total current so that the pulser is only called on to produce $\frac{1}{4}$ per cent of the mean beam power instead of the whole beam power which is the case with the usual arrangement. We shall return to the discussion of some special-purpose guns later.

4. THE PIERCE DESIGN METHOD

We now outline Pierce's design method, Pierce[(3, 8)]. The requirement is to achieve a space-charge limited flow from the cathode to the anode. The electrode trajectories are to be rectilinear in the cathode–anode space giving rise to a laminar flow beam. In the supposedly field-free space beyond the anode aperture the beam will diverge under the influence of (a) the electrostatic lens effect of the anode aperture and (b) the influence of space-charge forces in the beam. Such flows can be achieved in planar, cylindrical and spherical systems and in the planar case it is quite easy to find analytical relations for the required electrode shapes. We now consider the planar geometry of Fig. 5.

The origin of coordinates is taken at one edge of the cathode. Inside the beam, Poisson's equation is

$$\frac{\mathrm{d}^2 V}{\mathrm{d}z^2} = -\rho/\varepsilon_0 = \frac{J_0}{\varepsilon_0 \sqrt{(2\eta V)}}, \tag{4}$$

where J_0 is the current density in amps per square metre. At the beam edge $y = 0$, we must have

$$\frac{\partial V}{\partial y} = \frac{\partial V}{\partial z} = 0, \tag{5}$$

$$V = f(z), \tag{6}$$

where $f(z)$ is the required solution of (4).

Outside the beam, Laplace's equation gives

$$\frac{\partial^2 V}{\partial y^2} + \frac{\partial^2 V}{\partial z^2} = 0 \tag{7}$$

and the solutions to (4) and (7) must be continuous across the

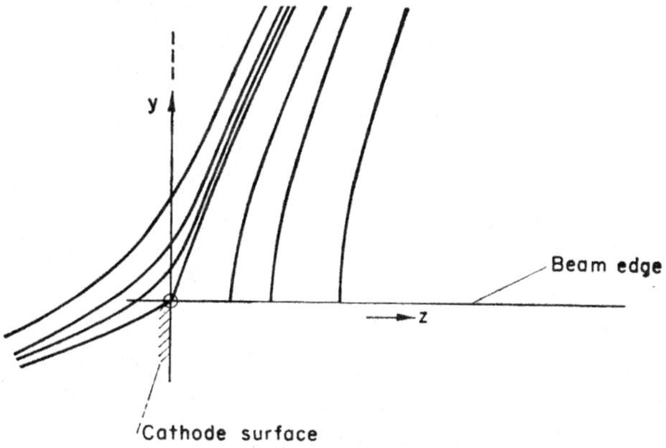

FIG. 5

boundary at $y = 0$, i.e. the solution to (7) is also subject to the boundary conditions (5) and (6). Now, a solution to Laplace's equation which is symmetrical about $y = 0$ and therefore for which $\partial V/\partial y = 0$ at $y = 0$, is

$$V = \tfrac{1}{2}[f(z+jy)+f(z-jy)]. \tag{8}$$

We next determine $f(z)$. If the initial velocities are ignored, we can use the boundary conditions $V = 0$, $\partial V/\partial z = 0$ at the cathode surface $z = 0$. These are the standard boundary conditions of Child's law and (4) can then be integrated twice to yield

$$V = \left[\frac{9J_0}{4\varepsilon_0(2\eta)^{1/2}}\right]^{2/3} z^{4/3} = 5.69\times 10^3 J_0^{2/3} z^{4/3}. \tag{9}$$

Then, according to (8), the potential variation outside the beam is

$$V = \frac{5.69 \times 10^3 J_0^{2/3}}{2} [(z+jy)^{4/3} + (z-jy)^{4/3}]. \qquad (10)$$

Equation (10) allows us to calculate the equipotentials which must be provided outside the beam to obtain the rectilinear field. Metal sheets are then bent so as to coincide with, for example, the 0 and V equipotentials and a rectilinear flow will result inside the beam.

5. NUMERICAL EXAMPLE

Suppose that we require an electron beam 2 cm deep, 0.2 cm broad, giving a current of 40 mA at a potential of 1 kV. Find the distance between the cathode and an anode at the full potential and the shape of the required electrodes.

The current density is 0.1 A/cm^2 or 10^3 A/m^2. Therefore, from (9),

$$z^{4/3} = \frac{1}{5.69 \times (10^3)^{2/3}} = 1.76 \times 10^{-3},$$

therefore

$$z = 0.85 \times 10^{-2}, \qquad m = 0.85 \text{ cm}.$$

Since the electrodes are to be the cathode and the anode at $V_0 = 1$ kV, the extension of the cathode electrode is a planar electrode making an angle of 67.5° with the edge of the beam, as shown in Fig. 5. The anode shape can be calculated, from (10), using $z = 0.85$ cm as the starting point at the edge of the beam. The shape is that shown in Fig. 5.

This gun has no area convergence. When convergence is desired, one must go over to cylindrical polar coordinates. Although the analytical solution for this case is not known, the necessary solution of Poisson's equation *inside* the flow has been computed by Langmuir and Blodgett(4). More important from the practical point of view is the case of converging flow between spherical caps, which they have also computed(4). The problem here is to find electrodes which will produce the required field *outside* the beam, so that the flow inside the beam is a rectilinear flow forming a conical frustrum of semi-angle θ. This flow is to be exactly the same as the flow observed if the cathode were extended to a sphere of radius r_c enclosing the anode sphere of radius r_a, if only the current in the cone were measured.

Langmuir and Blodgett determine the current in terms of a tabulated parameter $(-\alpha)^2$ which is a function of (r_c/r) (Table 1). The total current is given by

$$I_0 = 14.67 \times 10^{-6}\, V_0^{3/2}\, \frac{1-\cos\theta}{(-\alpha_a)^2}, \tag{11}$$

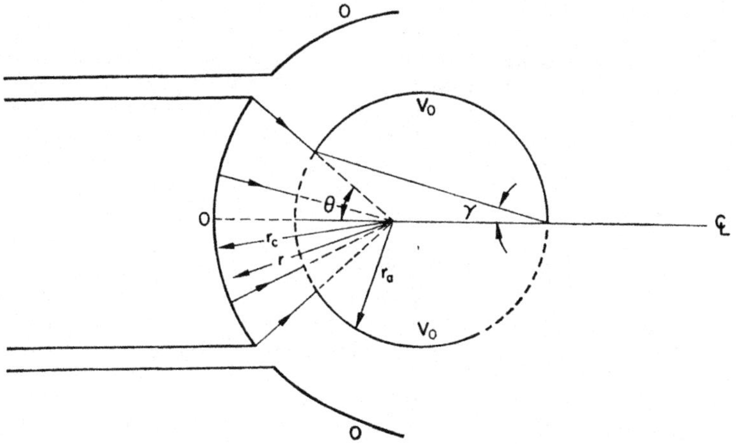

FIG. 6

where $(-\alpha_a)^2$ is the value of $(-\alpha)^2$ approximate to the anode, i.e. to (r_c/r_a). The voltage at any other radius r then follows from

$$V_{(r)}^{3/2} = \frac{I_0(-\alpha_r)^2 \times 10^6}{14.67\,(1-\cos\theta)}. \tag{12}$$

To use this result to find the electrode shapes for a given perveance and value of θ, we use eq. (12) to calculate the potentials at several values of r in the cathode–anode space. A model is then set up in an electrolytic trough (Fig. 7), usually the type with a tilted bottom (Bowman-Manifold and Nicoll[9], Beck[10]), and a strip of insulating material is adjusted to delineate the edge of the beam at angle θ. Since no current enters or leaves the electrolyte at the strip, there can be no normal potential gradient there and this boundary condition is automatically fulfilled. Several contacts are mounted on the strip at the radii previously used in calculating the voltages, and the electrodes outside the strip are bent until the potentials on the

TABLE 1. LANGMUIR'S α^2 FOR SPHERICAL DIODES

(r_c = radius of emitter; r = radius at any point P; α^2 applies to case where P is outside emitter; $r > r_c$; $(-\alpha)^2$ applies to case where P is inside emitter, $r_c > r$.)

r/r_c or r_c/r	α^2	$(-\alpha)^2$	r/r_c or r_c/r	α^2	$(-\alpha)^2$
1.0	0.0000	0.0000	6.5	1.385	13.35
1.05	0.0023	0.0024	7.0	1.453	15.35
1.1	0.0086	0.0096	7.5	1.516	17.44
1.15	0.0180	0.0213	8.0	1.575	19.62
1.2	0.0299	0.0372	8.5	1.630	21.89
1.25	0.0437	0.0571	9.0	1.682	24.25
1.3	0.0591	0.0809	9.5	1.731	26.68
1.35	0.0756	0.1084	10	1.777	29.19
1.4	0.0931	0.1396	12	1.938	39.98
1.45	0.1114	0.1740	14	2.073	51.86
1.5	1.1302	0.2118	16	2.189	64.74
1.6	0.1688	0.2968	18	2.289	78.56
1.7	0.208	0.394	20	2.378	93.24
1.8	0.248	0.502	30	2.713	178.2
1.9	0.287	0.621	40	2.944	279.6
2.0	0.326	0.750	50	3.120	395.3
2.1	0.364	0.888	60	3.261	523.6
2.2	0.402	1.036	70	3.380	663.3
2.3	0.438	1.193	80	3.482	813.7
2.4	0.474	1.358	90	3.572	974.1
2.5	0.509	1.531	100	3.652	1144
2.6	0.543	1.712	120	3.788	1509
2.7	0.576	1.901	140	3.903	1907
2.8	0.608	2.098	160	4.002	2333
2.9	0.639	2.302	180	4.089	2790
3.0	0.669	2.512	200	4.166	3270
3.2	0.727	2.954	250	4.329	4582
3.4	0.783	3.421	300	4.462	6031
3.6	0.836	3.913	350	4.573	7610
3.8	0.886	4.429	400	4.669	9303
4.0	0.934	4.968	500	4.829	13,015
4.2	0.979	5.528	600	4.960	
4.4	1.022	6.109	800	5.165	
4.6	1.063	6.712	1000	5.324	
4.8	1.103	7.334	1500	5.610	
5.0	1.141	7.976	2000	5.812	
5.2	1.178	8.636	5000	6.453	
5.4	1.213	9.315	10,000	6.933	
5.6	1.247	10.01	30,000	7.693	
5.8	1.280	10.73	100,000	8.523	
6.0	1.311	11.46			

contacts approach the calculated potentials to the required degree of accuracy. Although there is only one set of electrodes which provides an exact solution to the problem, an infinite set of approximations exists. These match the required potential distribution more or less accurately and the choice between them is made on the basis of convenience. Spangenberg, Helm and Field[(11)] give several families of electrodes found in this way for various angles of convergence.

The straightforward technique described above works well for perveances of up to about 0.5×10^{-6}, but beyond this figure various

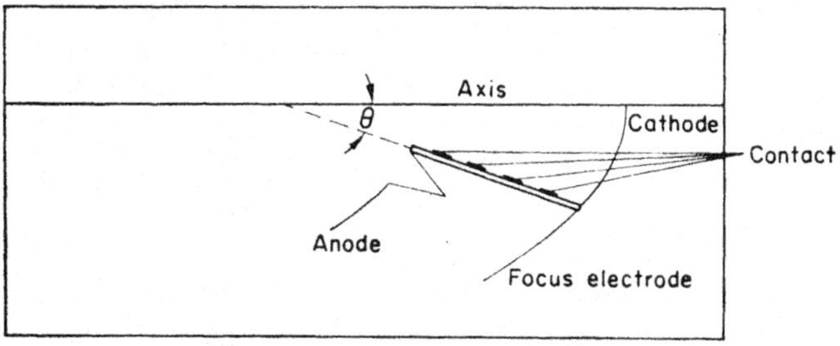

FIG. 7

modifications begin to be required. Let us first discuss the effect of the anode aperture.

6. ANODE APERTURE EFFECTS

We assume that the gun is used as a diode, which is usually the case in present-day valves. The first anode effect we have then to consider is the lens effect, which occurs when electrons pass through an aperture between spaces in which fields of different magnitudes are present. Davisson and Calbick[(12)] gave the following expression for the focal length of the lens formed by a circular aperture in a plane at potential V_0 separating fields of strength E_l and E_r (left and right) for electrons moving to the right. The focal length is

$$f = \frac{4V_0}{E_l - E_r}$$

in the case of our gun $E_r = 0$, so

$$f = \frac{4V_0}{E_a}, \tag{13}$$

where E_a is the field strength at anode surface and is equal to $-(\partial V/\partial r)r = a$.

This lens will diverge the beam so that it approaches the axis at angle γ (Fig. 6) rather than at the original angle θ. From the lens laws

$$\sin \gamma = \sin \theta \left(1 - \frac{r_a}{f}\right). \tag{14}$$

We can thus readily calculate the direction of the electrons after emerging from this diverging lens. If the anode–cathode spacing is too small, the beam will diverge from the aperture. This happens when $r_a/r_c = 0.71$, for in this case $E_a = 4V_0/r_a$.

However, the major importance of the anode aperture is due to another effect. The radius of the anode aperture is $r_a \sin \theta$. The potentials inside the beam are calculated for a solid anode. For the apertured anode the potential on the axis of the beam at a distance $r_c - r_a$ from the cathode will be lower than the assumed potential. This perturbation of potential will extend to a distance of about $2r_a \sin \theta$ on either side of the intercept which the anode radius vector r_a makes with the beam axis. When $2r_a \sin \theta$ is a considerable fraction ($>\frac{1}{2}$) of the anode–cathode spacing $r_c - r_a$, the perturbation of the field will certainly be sufficient to decrease the beam current below the calculated value. Curves showing the perveance and the locus of points for which the anode aperture diameter $= (r_c - r_a)$ and $\frac{1}{2}(r_c - r_a)$ have been prepared by Samuel[13]. These show that the perturbation reaches half-way to the cathode for perveances greater than 2×10^{-6}.

Pierce[8], Huber[14] (see also Beck[10], p. 164) and others have considered the matching of Pierce guns to space-charge flows in tunnels. We shall deal with this question later, after we have discussed such flows, but it is worth noting that Huber was able to show that when the gun was optimally matched to the tunnel $r_c/r_a = 2.15$ and the ratio of maximum beam current density to cathode current density was 26:1. Practical high perveance guns usually give smaller ratios than this.

We now turn to the question of calculating the behaviour of the beam during its passage through the valve proper, when it is in the quasi-cylindrical state.

7. ELECTRON BEAMS IN FIELD FREE SPACE

The simplest type of beam is that which emerges through the aperture in the gun anode into space in which there are neither electrostatic nor magnetic fields. As we shall soon see, the beam itself will set up electrostatic fields and our problem is to calculate the electron trajectories taking these fields into consideration. They would, of course, be neglected in classical electron optics.

A beam entering a cylindrical tunnel

The arrangement and notation for this problem are shown in Fig. 8. Several assumptions will be made and it is convenient to discuss these in a group.

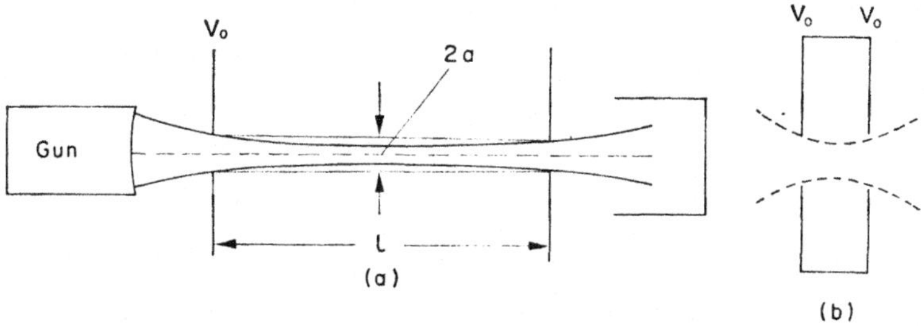

FIG. 8

1. The tunnel length is assumed to be much greater than its diameter, i.e. $d/l \ll 1$. This ensures that the regions of entry to and exit from the tunnel where there is some penetration of electrostatic field from the gun/collector spaces form a minor part of the total beam path and the perturbations resulting from these fields can be neglected. This assumption also ensures that only transverse potential variations are important. If the arrangement were that of Fig. 8b, the longitudinal field variation would have to be included.

2. The beam is assumed to exhibit laminar flow, that is, the electron trajectories never cross one another. If a cone of electrons of semi-angle θ_0 (Fig. 9) is converging to a point distant l from the plane where the cone has a cross-sectional radius of r_0, then the electrons at a smaller radius r_1 are moving towards the axis at an angle θ, where

$$\frac{\tan\theta_1}{\tan\theta_0} \doteqdot \frac{\theta_1}{\theta_0} = \frac{r_1}{r_0}. \tag{15}$$

This behaviour would be expected in a perfectly designed and made Pierce gun and it is a very useful assumption. There are, however, good reasons for believing that many useful electron beams do not, in fact, exhibit laminar flow, and we shall examine this question in more detail later on.

3. Let us assume that the initial radially inward velocities are small so that the axial electron velocity u_z, which really varies by a small amount, may be assumed constant, and that its constant value is the velocity equivalent to the wall potential V_0.

Now let us discuss the physical behaviour of the beam after it enters the tunnel at the anode plane. At this plane the electrons are travelling in a cone towards the axis at an angle which is the angle of convergence of the gun modified by the divergence of the aperture lens [eq. (14)]. In the tunnel the space-charge forces, which were

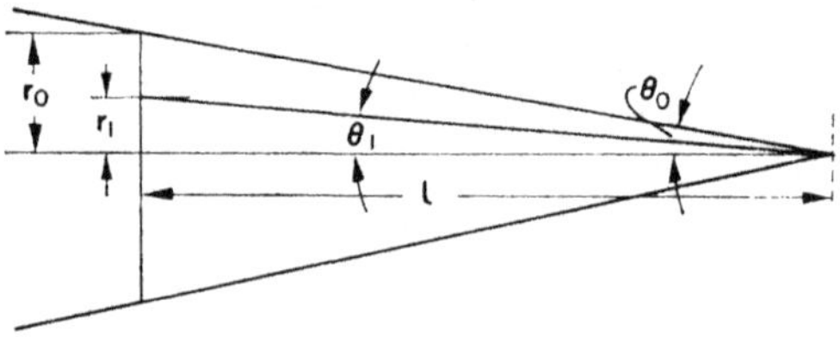

FIG. 9

compensated by the fields in the region outside the tunnel, give rise to a radial electric field and therefore to a radial force on the electrons which, as we shall see in a moment, depends inversely on the beam radius. The radial force reduces the initial inwards velocity, which eventually becomes zero at some plane where all the electrons instantaneously travel parallel to the axis. To the right of this plane the radial space-charge field will cause the beam to diverge. The beam profile is symmetrical about the plane of the throat. We can now produce the argument which shows that the tunnel will pass a maximum current when the plane of zero radial velocity is placed at the centre of the tunnel. This plane will, in future, be called the "throat".

First, suppose that a low current beam enters the tunnel, grazing the walls at entry, and forms a throat at a plane to the right of the tunnel centre. Then, increasing the beam current while maintaining the diameter of the entrance pupil constant will increase the space-charge forces and the throat will move nearer to the tunnel centre. When the centre is reached, the symmetry ensures that the beam which grazes at entry also grazes at exit. Any further increase in beam current will cause beam interception before the exit plane is reached and the current available beyond the tunnel will be reduced.

To calculate the values we start from the throat plane (all electron trajectories parallel with axis) and compute the trajectories from there using Gauss's theorem to find the radial field. Let b = beam radius at throat, ρ = space-charge density.

Applying Gauss's theorem at the throat,

$$2\pi b \,.\, E_r = \pi b^2 \rho/\varepsilon_0,$$

$$\therefore\; E_r = \frac{\rho b}{2\varepsilon_0}. \tag{16}$$

But

$$\rho = \frac{-I_0}{\pi b^2 u_z} \tag{17}$$

and

$$u_z = u_0 \qquad \text{(assumption 3)}$$

$$= \sqrt{(2\eta V_0)} \tag{18}$$

with

$$\mathrm{d}z = u_0\,\mathrm{d}t. \tag{19}$$

Therefore the equation for the trajectory of the outermost electron is

$$m\ddot{b} = eE_r$$

or

$$\frac{\mathrm{d}^2 b}{\mathrm{d}z^2} = \frac{1}{4\pi\varepsilon_0\sqrt{(2\eta)}} \cdot \frac{I_0}{V_0^{3/2}} \cdot \frac{1}{b}. \tag{20}$$

This can be integrated, multiplying both sides by $2(\mathrm{d}b/\mathrm{d}z)$ to

$$\left(\frac{\mathrm{d}b}{\mathrm{d}z}\right)^2 = \frac{1}{2\pi\varepsilon_0\sqrt{(2\eta)}} \cdot \frac{I_0}{V_0^{3/2}} \ln{(r/b)} \tag{21}$$

$$= \frac{K}{2} \ln{(r/b)}, \tag{22}$$

where

$$K = \frac{1}{\pi\varepsilon_0\sqrt{(2\eta)}} \cdot \frac{I_0}{V_0^{3/2}} \tag{23}$$

$$= 6.04 \times 10^4\, I_0/V_0^{3/2}. \tag{24}$$

Thus

$$z = \int_b^r \frac{\mathrm{d}b}{(K/2 \ln r/b)^{1/2}}. \tag{25}$$

By changing variables to

$$R = r/b, \qquad Z = \left(\frac{K}{2}\right)^{1/2} \frac{z}{b},$$

this becomes

$$Z = \int_1^R \frac{\mathrm{d}R}{(\ln R)^{1/2}}. \tag{26}$$

This equation has been discussed by many authors starting with Watson[15], Thompson and Headrick[16] and Pierce[8]. Pierce reduces the integral to a form which can be integrated using values of the function

$$F(x) = e^{-x^2} \int_0^x e^{y^2} \, \mathrm{d}y.$$

tabulated by Miller and Gordon[17]. The integration results in the universal beam spread curve (Fig. 10). To find the maximum current we consider the point ($R = a/b$, $Z = \sqrt{(K/2)}.l/2b$) which represents the exit conditions. We know neither b nor K. For a fixed K, however, this point lies on a straight line of slope $a/l\sqrt{(2/K)}$ going through the origin. But the point must also lie on the beam spread curve so that K must be adjusted until the straight line through the origin becomes tangent to the curve. This happens at $R = 2.4$, $Z = 2.65$.

Then

$$b = \frac{a}{2.4}, \tag{27}$$

$$2.65 = \left(\frac{K}{2}\right)^{1/2} \frac{l}{2b} = \left(\frac{K}{2}\right)^{1/2} \frac{1.2\, l}{a}. \tag{28}$$

Putting in the numerical value of K, we find

$$I_{\max} = 38.9 \times 10^{-6}\, V_0^{3/2} \left(\frac{2a}{l}\right)^2. \tag{29}$$

To summarize, we must arrange our gun so that the throat is found at the centre of the tunnel, where its diameter is 1/2.4 times the tunnel diameter and at the same time achieve a gun perveance of $38.9(2a/l)^2 \times 10^{-6}$. We note that the conditions to achieve the maximum current just discussed also require the real beam to be shaped so that the tangent to the trajectory of the outermost electron, at the point of entry to the tunnel, must go through the centre of the tunnel. Thus the angle γ as defined by eq. (14) must also be given by

$$\tan \gamma = \left(\frac{2a}{l}\right). \tag{30}$$

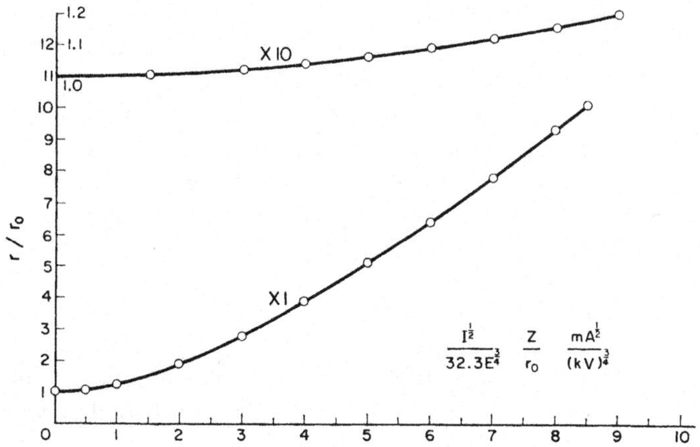

Universal beam spread curve

FIG. 10

Since our assumptions make γ small, we can put $\tan \gamma \fallingdotseq \sin \gamma$ and obtain

$$\left(\frac{2a}{l}\right) = \sin \theta \, (1 - r_a/f). \tag{31}$$

This equation is the basic relation which matches the trajectories in the gun region of those required in the tunnel. Its implications are further discussed in refs. 10 and 14.

8. A BEAM IN A RECTANGULAR TUNNEL

Exactly the same method can be used to discuss the motion of an electron beam focused by cylindrical lenses into a long rectangular tunnel. This situation can be pictured by considering Fig. 8a as a section on any horizontal plane through the device. The calculation here involves only straightforward analytical integration, and full details are given in ref. 10. The result is

$$I_{max} = 83.3 \times 10^{-6} \frac{th}{l^2} V_0^{3/2}, \tag{32}$$

where t is the tunnel width and h is the tunnel height.

In this case, the outermost electron trajectories are parabolic and they become tangent to the axis at the centre of the system. The beam has therefore zero thickness at the throat. We know that this will not, in fact, be the case because of the thermal limitations and closer investigation shows that thermal velocities limit the current to a value less than that given by eq. (32). We shall take up this question later. However, continuing with the simple analysis, we find that, from the geometry of the parabola, the tangent to the trajectory at entry must cross the axis at a distance $l/4$ from the plane of entry. In this case then,

$$\tan \gamma = \frac{4t}{l}. \tag{33}$$

Matching is achieved as before using, of course, the cylindrical lens formula and the values of field valid for cylindrical space-charge flows.

Electrostatic focusing of the type described in these two sections is very simple and convenient and the system was used in early klystrons and still is used in nearly all reflex klystrons. However, if $2a/l = 1/6$ and $V_0 = 1$ kV the maximum current is only about 34 mA and the perveance is barely over unity. In typical multi-cavity klystrons and travelling-wave tubes $2a/l$ is much smaller than 1/6, perhaps 1/20–1/50, and the perveance available from this type of flow is much too low to be useful. The perveance limitation inherent in single potential electrostatic focusing can be overcome by the addition of a strong magnetic field along the axis of the valve. When this is done, the radial velocity component due to the space-charge forces gives rise to a magnetic force along the axis and the electron trajectories are forced into tight helices. The beam is largely prevented from spreading but the space-charge field now gives a potential distribution across the beam. With large beam currents this potential distribution can result in considerable diminution of the axial potential. The beam is no longer a uni-velocity beam and the d.c. distribution of velocities may be large enough to influence (adversely) the operation of velocity modulation devices.

9. MAGNETIC CONFINEMENT OF AN ELECTRON BEAM

With the same assumptions as before we can apply ourselves to calculating the current passed by a circular tunnel when the system is immersed in a very strong field B acting in the direction shown in Fig. 11. We are only concerned with radial fields as the system is

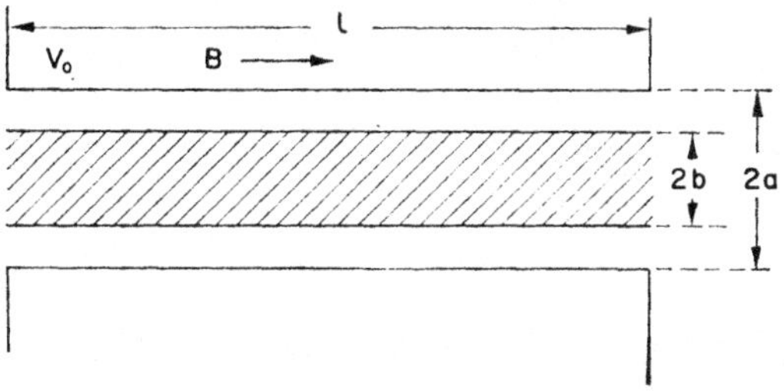

FIG. 11

axially symmetrical. Then, outside the beam $b<r<a$ where Laplace's equation holds we have

$$\frac{\partial^2 V_2}{\partial r^2}+\frac{1}{r}\frac{\partial V_2}{\partial r}=0. \tag{34}$$

Inside the beam $r<b$, we have Poisson's equation

$$\frac{\partial^2 V_1}{\partial r^2}+\frac{1}{r}\frac{\partial V_1}{\partial r}=\rho/\varepsilon_0 \tag{35}$$

together with

$$\rho=\frac{-I_0}{\pi a^2 u_0}, \qquad u_r=\sqrt{(2\eta V_r)}.$$

Here u_r, V_r are the velocity and potential at radius r.

The boundary conditions are that V and E_r should be continuous at $r=b$, that is

$$V_1=V_2, \qquad \frac{dV_1}{dr}=\frac{dV_2}{dr} \qquad \text{at } r=b \tag{36}$$

and

$$V_2=V_0 \qquad \text{at } r=a. \tag{37}$$

Introduce dimensionless coordinates

$$\Phi_n = \frac{V_n}{V_m}, \quad R = r\left[\frac{I_0}{\pi b^2 \varepsilon_0 (2\eta)^{1/2}}\right]^{1/2} V_m^{-3/4} = \frac{r}{b}(K)^{1/2}\left(\frac{V_0}{V_m}\right)^{3/4}. \tag{38}$$

K was defined earlier in eqs. (23) and (24) and V_m is the potential on the axis. Then

$$\frac{d^2\Phi_1}{dR^2} + \frac{1}{R}\frac{d\Phi_1}{dR} = \Phi^{-1/2}, \tag{39}$$

$$\frac{d^2\Phi_2}{dR^2} + \frac{1}{R}\frac{d\Phi_2}{dR} = 0. \tag{40}$$

At $r = b$, we have

$$R_b = K^{1/2}\left(\frac{V_0}{V_m}\right)^{3/4} = K^{1/2}\Phi_0^{3/4} \tag{41}$$

and at $r = a$,

$$R_a = a/b\,(K)^{1/2}\,\Phi_0^{3/4}. \tag{42}$$

Equations (39) and (40) can now be integrated graphically or numerically subject to the boundary conditions (41) and (42). This has been done by several authors including Smith and Hartmann(18), Guénard(19) and Brewer(20). Here, we only consider the case when the beam fills the tunnel, i.e. $a/b = 1$. Guénard gives a curve of Φ versus R for this case. Maximum current occurs when the value of R given by eq. (41) falls on the curve. This gives

$$R = K^{1/2}\,\Phi^{3/4}, \quad \frac{dR}{d\Phi} = 3/4\,K^{1/2}\,\Phi^{-1/4},$$

therefore

$$\Phi = 3/4\,R\,\frac{d\Phi}{dR}. \tag{43}$$

Figure 12 shows the intersection of the $\Phi \sim R$ and $\Phi \sim 3/4\ R\Phi'$ curves to be at $R = 5.12$, $\Phi = 5.65$. Putting in these values we find

$$I_{\max} = 32.5 \times 10^{-6}\,V_0^{3/2}. \tag{44}$$

Thus, for a beam filling the tunnel, the perveance is independent of the tunnel length or diameter and has the high value of 32.5×10^{-6}. Clearly, the axial potential is very low, only $V_0/5.65$ or 17.7 per cent of the wall potential when the full current is flowing. This would mean that the electron velocity on the axis is less than half the value on the periphery. This would be intolerable in any velocity modulation device so much smaller perveances have to be used if the potential depression is to be kept to more reasonable values. A potential depression of 10 per cent, corresponding with a velocity depression of 5 per cent, occurs when the perveance is about 6×10^{-6}. Even this figure is much greater than we found for the electrostatic case.

The treatment thus far is, however, very schematic. It makes no reference to the way in which the electrons enter the field or the way in which they originate in the gun. Without this knowledge it is not

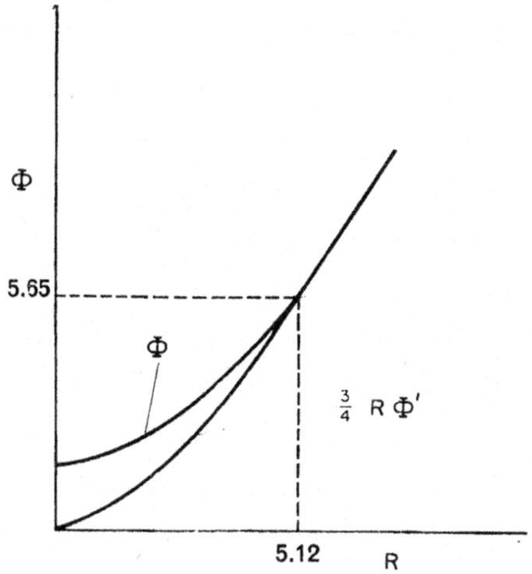

FIG. 12

possible to calculate the value of magnetic field required to obtain an acceptable approximation to a cylindrical beam. We shall therefore now discuss these questions in more detail, starting with a discussion of a case in which a truly cylindrical beam can be achieved, at least in theory.

10. BRILLOUIN FOCUSING OF A CYLINDRICAL BEAM

Brillouin[21] gives the following analysis for the conditions required to achieve a cylindrical beam. The current density, electric and magnetic fields, etc., are all considered to be uniform in the θ direction.

The Lorentz force law can then be written in cylindrical polar co-ordinates as

$$\ddot{r} - r\dot{\theta}^2 = -\eta (E_r + B_z r\dot{\theta}), \tag{45}$$

$$1/r \, \mathrm{d}/\mathrm{d}t (r^2\dot{\theta}) = -\eta (-B_z \dot{r} + B_r \dot{z}), \tag{46}$$

$$\ddot{z} = -\eta (E_z - B_r r\dot{\theta}). \tag{47}$$

The magnetic field, from one of Maxwell's equations, is given by $\nabla . \boldsymbol{B} = 0$, which when $B_\theta = 0$ can be expanded to give

$$B_r = -r/2 \left(\frac{\partial B_z}{\partial z}\right)_{r=0}. \tag{48}$$

Thus the field is known once the variation of B_z along the z axis is determined.

Also,

$$\mathrm{d}/\mathrm{d}t = \dot{r}\,\partial/\partial r + \dot{z}\,\partial/\partial z$$

and therefore eq. (46) can be integrated to give

$$r^2\dot{\theta} = \eta \int \left(B_z r\dot{r} + \frac{r^2 \dot{z}}{2} \frac{\partial B_z}{\partial z}\right) \mathrm{d}t = \eta \frac{B_z r^2}{2} + C. \tag{49}$$

On the cathode surface, $r = r_0$, $\dot{\theta} = 0$ and $B_z = B_0$. Therefore

$$r^2\dot{\theta} = \eta/2 (B_z r^2 - B_0 r_0^2). \tag{50}$$

Equation (50) is called Busch's theorem. The angular frequency $\eta B_z/2$ is often called the Larmor frequency, symbol ω_L. If we introduce this and also put $\eta B_0/2 = \omega_0$, eq. (50) simplifies to

$$\dot{\theta} = \left(\omega_L - \omega_0 \frac{r_0^2}{r^2}\right). \tag{51}$$

With this substitution eq. (45) becomes

$$\ddot{r} = -\eta E_r + r\left(\omega_0^2 \frac{r_0^4}{r^4} - \omega_L^2\right). \tag{52}$$

But, from Gauss's theorem for a uniform cylinder of current I_0

$$E_r = \frac{-I_0}{2\pi\varepsilon_0 u_0 r}$$

so that

$$\ddot{r} = \frac{\eta I_0}{2\pi\varepsilon_0 u_0 r} + r\left(\omega_0^2 \frac{r_0^4}{r^4} - \omega_L^2\right). \tag{53}$$

From eq. (53) we observe that the magnetic field, now measured by ω_L, is most effective in overcoming the diverging effects of space charge when $\omega_0 = 0$, that is, when the cathode is located inside a perfect magnetic screen so that $B = 0$ on the cathode surface. If this is the case, we can make $\ddot{r} = 0$ by suitable adjustment of ω_L. The required value is

$$\omega_L^2 = \frac{\eta I_0}{2\pi\varepsilon_0 u_0 r^2} \tag{54}$$

or

$$B_B^2 = \frac{2I_0}{\pi\eta\varepsilon_0 u_0 r^2} = \frac{\sqrt{2}\, I_0}{\pi\eta^{3/2}\,\varepsilon_0 r^2\, V_a^{1/2}}. \tag{55}$$

Numerically,

$$B_B^2 = 7.0 \times 10^{-7}\,\frac{I_0}{V_a^{1/2}\, r^2}. \tag{56}$$

From eq. (51) we see that $\dot{\theta} = \omega_L$ when $\omega_0 = 0$ and from eq. (52)

$$-\eta E_r = \omega_L^2 r. \tag{57}$$

Therefore

$$\partial V/\partial r = \frac{r\omega_L^2}{\eta}$$

or

$$V = V_a + \frac{r^2\omega_L^2}{2\eta}. \tag{58}$$

In eqs. (56) and (58) V_a is the potential on the axis of the beam. We shall justify the use of V_a in eq. (56) very shortly. We can equate the potential energy to the sum of the translational and rotational kinetic energies and obtain

$$(\dot{z})^2 + (r\dot{\theta})^2 = 2\eta V. \tag{59}$$

Therefore

$$(\dot{z})^2 + r^2\omega_L^2 = 2\eta\left(V_a + \frac{r^2\omega_L^2}{2\eta}\right);$$

therefore

$$\dot{z} = \sqrt{(2\eta V_a)}. \tag{60}$$

This shows that all the electrons have the same axial velocity, a velocity which is, however, not that corresponding with the potential of the surrounding conductors but is instead the potential on the beam axis. We can now substitute for ω_L^2 in eq. (58) to find I_0. The result is

$$I_0 = \frac{\pi\varepsilon_0\eta^{3/2}\, B_B^2 r^2}{\sqrt{2}}\left(V_0 - \eta\,\frac{B_B^2 r^2}{8}\right). \tag{61}$$

The maximum current and perveance can be found from maximizing this, when we find

$$\eta\,\frac{\hat{B}_B^2 r^2}{8} = \frac{2V_0}{3}, \tag{62}$$

$$\hat{I}_0 = \frac{16}{3\sqrt{6}}\,\pi\varepsilon_0\eta^{1/2}\,V_0^{3/2}, \tag{63}$$

$$\hat{p} = 25.4\times 10^{-6}. \tag{64}$$

Therefore, if the injected current is given the maximum value of $I_0 = 25.4\times 10^{-6}\ V_0^{3/2}$ amps, the potential on the axis is depressed to $1/3\,V_0$ and the velocity inside the beam is rather below 50 per cent of the velocity corresponding with the applied potential. However, in this case all the electrons have the same axial velocity so that we are not troubled by a velocity distribution. If it were not for high frequency instabilities (Beck[(22)]), the Brillouin beam could be used at full perveance in microwave devices if the coupling factors, etc., were calculated for an accelerating voltage equal to a third of the voltage actually applied to the valve.

Brillouin beams are frequently used today, but the perveance of existing guns limits the beam perveance to about 10 per cent of the maximum value. Under these conditions the axial velocity is not depressed very much below the maximum available velocity. The potential for any current below the maximum can be calculated as follows. If we divide eq. (61) by eq. (63) and eliminate V_0 by eq. (62), we obtain

$$\frac{I_0}{\hat{I}_0} = \left(\frac{B_B}{\hat{B}_B}\right)^2\left\{3-2\left(\frac{B_B}{\hat{B}_B}\right)^2\right\}^{1/2}$$

or

$$y = x^2(3-2x^2)^{1/2}. \tag{65}$$

From eq. (56)

$$\left(\frac{B_B}{\hat{B}_B}\right)^2 = \frac{I_0\,V_0^{1/2}}{\sqrt{3}\,\hat{I}_0\,V_a^{1/2}}$$

or

$$x^2 = y/\sqrt{3}\,(V_0/V_a)^{1/2},$$

i.e.

$$V_a = \frac{V_0}{3}(y/x^2)^2. \tag{66}$$

Figure 13 shows plots of eq. (65) and of $1/3(y/x^2)^2$ which allow one

$\frac{B_B}{\hat{B}_B}$	$\frac{I_0}{\hat{I}_0}$	$\frac{V_a}{V_0}$
0.1	0.017	0.993
0.2	0.068	0.974
0.3	0.151	0.940
0.4	0.261	0.894
0.5	0.395	0.834
0.6	0.543	0.760
0.7	0.695	0.673
0.8	0.838	0.573
0.9	0.951	0.460
1.0	1.000	0.333

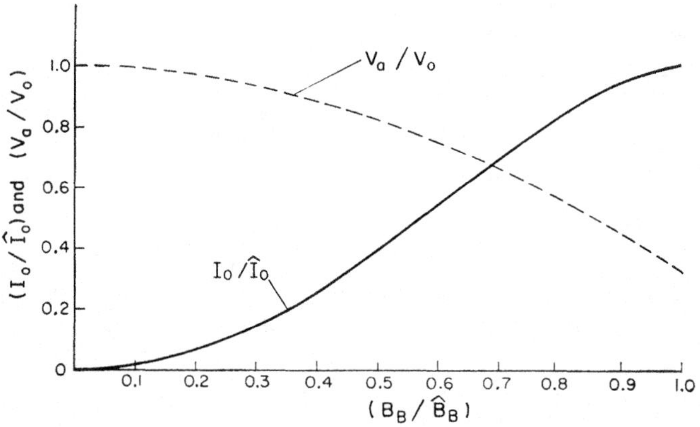

FIG. 13

to determine B_B and V_a for any value of I_0 since $\hat{I}_0$ and $\hat{B}_B$ are readily calculable.

Practical applications of Brillouin flow face the difficulty that the magnetic shield must be apertured to allow the exit of the beam. The magnetic field thus cannot jump abruptly to the desired value but increases gradually over an axial distance of about three aperture diameters. This gradual increase in field constitutes a magnetic lens whose effect on the beam must be considered. To do this, assume that the beam diameter is nearly constant in the region of varying field. Then, the arguments used to deduce the magnitude of the final field (for the current and voltage considered) are based on equal space charge and magnetic forces at exit from the lens. In the lens region the space-charge forces will be greater than the magnetic forces at the plane considered. The beam must therefore diverge under the influence of space charge while passing through the lens. This divergence can be overcome if the beam is given an initial convergence at entry to the lens. Müller[(23)] has analysed the system and shows that the beam should converge so that, in the absence of magnetic field, the throat is located at the plane where $B = 0.7\, B_B$.

In practical Brillouin guns it is found that the field required for optimum beam efficiency (ratio of current passing through the electrode system to total current entering the system) is 1.5–2.0 B_B. The large perturbations produced in a Brillouin beam by r.f. fields have already been mentioned[(22)]. It has also been found that electron beams emerging from Brillouin guns are very noisy. The effect is discussed in detail by Rigrod[(24)]. Finally, recent work by Harker[(25)], Johnson[(26)], Ashkin[(27)] and Hermann[(28)] calls the whole concept of laminar flow from magnetically shielded guns into question. We shall return to this work later on. For these reasons other types of magnetic focusing are being used in spite of their less efficient utilization of magnetic field.

11. FOCUSING IN CONFINED FLOW

Two versions of confined flow are commonly used. In the first of these, a cathode of diameter substantially equal to the beam diameter is completely immersed in a purely longitudinal magnetic field which is made strong enough to restrict the transverse motion to an acceptable amount. The second form is that in which the lines of magnetic field in the gun region are designed to coincide as closely as possible with the electron trajectories in that region. The first type is obviously limited to cases in which the beam current density is no greater than the working current density of the cathode. In the second, the normal convergence ratio of the gun can be achieved.

We now consider case 1.

The magnetic field now has the constant value B_0 along the whole length of the beam. Equation (53) then becomes

$$\ddot{r} = \eta\, \partial V/\partial r + r\omega_0^2\left(\frac{r_0^4}{r^4}-1\right). \tag{67}$$

The developed paths of the peripheral electrons in a longitudinal magnetic field must be helices, so the surface of the beam will be scalloped. At the equilibrium radius of the beam r_m, $\ddot{r}=0$. We may also consider that, if the scalloping is small, as it must be for a practically useful beam, we can estimate the radial field by using Gauss's theorem for a cylinder of radius r_m. Using these substitutions in eq. (67) we find

$$1-\left(\frac{r_0}{r_m}\right)^4 = \frac{\sqrt{2}\, I_0}{\pi\varepsilon_0 \eta^{3/2} B_0^2\, V^{1/2}\, r_m^2}.$$

Following Pierce[8] we put

$$k = \frac{I_0}{\sqrt{2}\,\pi\varepsilon_0\eta^{3/2} B_0^2\, V^{1/2}\, r_0^2} = \frac{3.5\times 10^{-7}\, I_0}{B_0^2\, V^{1/2}\, r_0^2} \tag{68}$$

and obtain

$$\left(\frac{r_m}{r_0}\right)^4 - 2k\left(\frac{r_m}{r_0}\right)^2 - 1 = 0$$

or

$$\frac{r_m}{r_0} = [(k^2+1)^{1/2}+k]. \tag{69}$$

If the beam is not to spread to a radius much greater than the cathode radius r_0, k must be small and

$$r_m \fallingdotseq r_0\,(1+k/2). \tag{70}$$

We next calculate the scalloping of the beam, using a perturbation method first applied by Kleen and Pöschl[29] to a convergent confined flow. We put

$$r = r_m(1+\delta), \tag{71}$$

where δ is assumed $\ll 1$, so that r^{-1} and r^{-3} can be represented by the first two terms of their binomial expansions. Substitution in (67) yields

$$\ddot{\delta} + 2\left(1+\frac{r_0^4}{r_m^4}\right)\omega_0^2\,\delta = 0, \tag{72}$$

i.e.

$$\delta = C_1 \cos \sqrt{\left[2\left(1+\frac{r_0^4}{r_m^4}\right)\right]}\,\omega_0 t + C_2 \sin \sqrt{\left[2\left(1+\frac{r_0^4}{r_m^4}\right)\right]}\,\omega_0 t. \quad (73)$$

At the cathode $t = 0$, $r = r_0$, therefore

$$\delta = \frac{r_m - r_0}{r_m} \cos \sqrt{\left[2\left(1+\frac{r_0^4}{r_m^4}\right)\right]}\,\omega_0 t. \quad (74)$$

The maximum beam diameter $r_m(1+\delta)$ then is equal to

$$r_m\left(2-\frac{r_0}{r_m}\right) \doteqdot r_m\left(1+\frac{k}{2}\right) = r_0\left(1+\frac{k}{2}\right)^2. \quad (75)$$

To keep the ratio of maximum to minimum beam diameter below 1.2, which is a practical figure, k must be <0.2. This, for small perveances, requires that $B_0 = 1.6\ \hat{B}_B$. The calculations, however, neglect the effects of diverging electrostatic lenses which may easily make matters worse.

12. CONVERGENT CONFINED FLOW

The system and the notation are shown in Fig. 14. The gun proper is located inside a magnetic shield whose diameter and position are adjusted so that the field lines coincide with the electron trajectories.

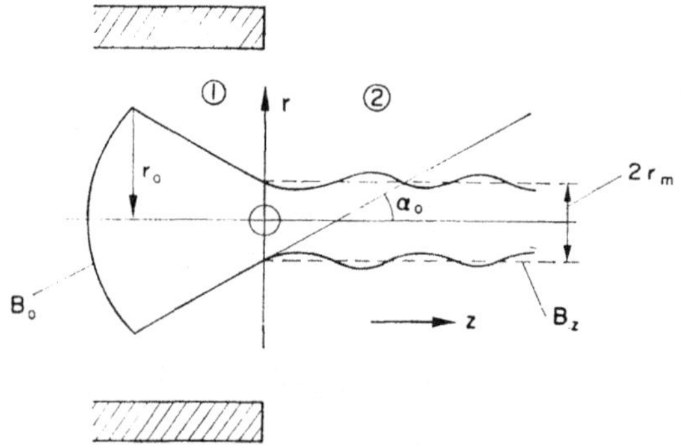

FIG. 14

The electron trajectories naturally have to be determined experimentally, either using a conventional electrolytic tank and applying the space-charge corrections by step-by-step methods, or using an automatic trajectory plotter with facilities for space-charge injection.

Once again, the theory starts with eq. (53) but B is now varying along the electron trajectories. After a little obvious algebra we can write

$$\frac{B_0^2}{B_z^2}\frac{r_0^4}{r_m^4}+\frac{2K}{r_m^2}-1=0, \tag{76}$$

where

$$K=kr_0^2=\frac{3.5\times10^{-7}I_0}{B_z^2V^{1/2}}. \tag{77}$$

Then

$$r_m=K^{1/2}\left[1+\sqrt{\left(1+\frac{B_0^2}{B_z^2}\frac{r_0^4}{K^2}\right)}\right]^{1/2}. \tag{78}$$

Put $r=r_m(1+\delta)$ as before, expand and we obtain

$$\ddot{\delta}+2(1+A)\,\omega_L^2\,\delta=0,$$

where

$$A=\left(\frac{B_0\,r_0^2}{B_z\,r_m^2}\right),\quad 0\leqslant A\leqslant 1. \tag{79}$$

Then

$$\delta=C_1\cos\sqrt{[2(1+A)]}\,\omega_L t+C_2\sin\sqrt{[2(1+A)]}\,\omega_L t. \tag{80}$$

In this case the origin is taken at the anode aperture separating region 1 from region 2. At the origin let $r=r_a$, and let the beam converge to the axis at angle α_0. Then

$$C_1=\frac{r_a-r_m}{r_m},\quad C_2=\frac{u_0\tan\alpha_0}{\sqrt{[2(1+A)]}\,\omega_L\,r_m}$$

and the maximum beam radius ratio is given by

$$\frac{r_{\max}}{r_{\min}}=1+\frac{r_a}{r_m}\sqrt{\left[\left(\frac{r_m}{r_a}-1\right)^2+\frac{1}{2(1+A)}\left(\frac{u_0\tan\alpha_0}{\omega_L\,r_a}\right)^2\right]}. \tag{81}$$

Figure 15 shows $r_{\max}/r_{\min}$ as a function of r_a/r_m with the second term under the radical as parameter. Good beam performance can be achieved by arranging matters so that $r_a/r_m \doteqdot 1.0$ so the electron beam should enter space free from electric fields at the desired mean diameter. Then, by increasing ω_L, the degree of ripple can be reduced to a predetermined value.

We note that if the electron trajectories exactly coincide with the B lines, then by definition $A = 1$. But, by Busch's theorem, this would ensure that $\dot{\theta} = 0$ for all the electrons and there would be no compensation of space-charge forces by magnetic forces. Convergent flow guns thus have to be finally adjusted by experiment, and a more exact theory would require the complete solution of the problem with the correct variation of the electric fields inserted in the gun region. This has not yet been achieved.

Convergent flow beams are much more stable under the influence of strong r.f. fields than are Brillouin beams, and for this reason they are becoming more common in modern microwave valves. However, the process of beam formation has not been studied nearly as closely as it has in Brillouin beams.

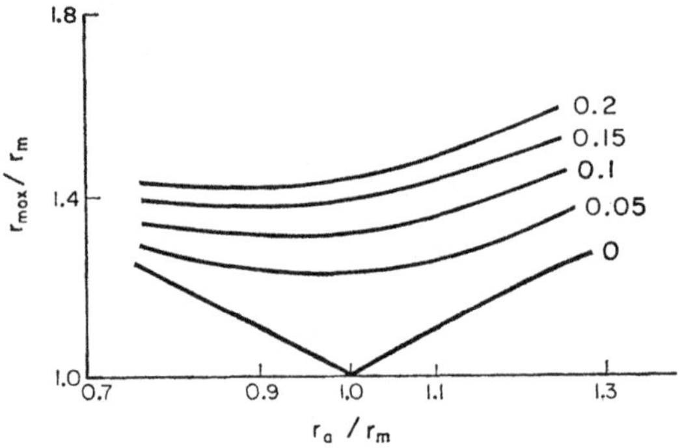

FIG. 15

Strong focusing

We now consider methods of maintaining a long electron beam by the use of a series of strong lenses, either electrostatic or magnetic, uniformly distributed along the length of the beam. If we consider for the moment the magnetic case, it can be shown by the following qualitative argument that the stray magnetic field can be much reduced and this will lead to a reduction in magnet weight. We suppose that the short lenses are arranged so that the direction of the longitudinal field is reversed on moving from each lens to the next one. Then, the stray field at a point distant by several lens diameters from any lens, the average field along the axis will be zero. Thus the stray field has been reduced. Mendel *et al.*[30], who first investigated this system, show that if N small magnets are used instead of one long one, the gain in weight is proportional to N^2. In practical arrangements gains of between N and N^2 have been obtained. Their focusing magnets demand that the field reversals should be sudden. Thus the magnets must be brought near to the beam. This has limited the system to use with valves such as travelling-wave tubes in which it is not necessary to have access to tuning mechanisms, etc., along the beam.

Recently, workers at the Mullard Research Laboratories have made a careful investigation of the special case in which only a single field reversal is used. This allows one to locate tuned klystron

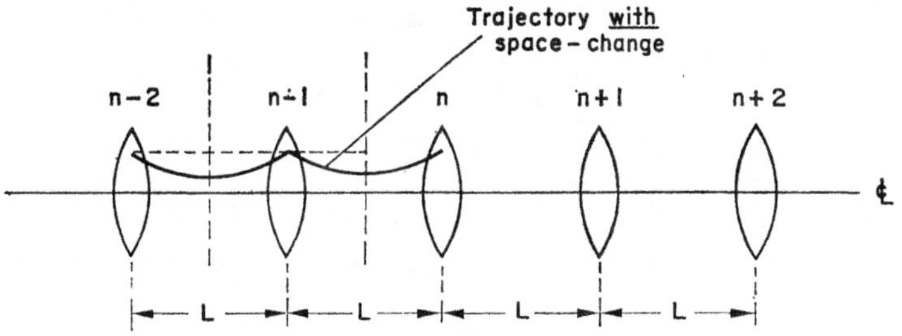

FIG. 16

cavities along the beam and at the same time represents a more economic use of magnetic material than does a single magnet[31].

Returning now to the general strong lens theory, we repeat the analysis of Pierce[8]. The lenses are arranged with uniform spacing L (Fig. 16) and have equal convergence C. Let the radius and slope of a paraxial ray, measured just to the right of the nth lens, be r_n and r'_n. Then

$$r_{n+1} = r_n + Lr'_n\,, \tag{82}$$

$$r'_{n+1} = r'_n - (r_{n+1}), \tag{83}$$

whence

$$r'_n = \frac{1}{L}(r_{n+1} - r_n), \tag{84}$$

therefore

$$r'_{n+1} = \frac{1}{L}(r_{n+2} - r_{n+1}), \tag{85}$$

and

$$r_{n+2}-(2-CL)\,r_{n+1}+r_n=0. \tag{86}$$

This difference equation has solutions of the form

$$r = A\cos n\theta + B\sin n\theta. \tag{87}$$

If we take $n=0$ and consider only the cosine term, the solution of (86) becomes

$$\theta = \cos^{-1}\left(1-\frac{CL}{2}\right) \tag{88}$$

and θ is then real for $CL<4$ and the electron trajectories follow a path which oscillates co-sinusoidally through the system. At the limiting convergence, $C=4/L$, the path of an electron which entered the first lens on a trajectory parallel to the axis would cross the axis at the centre of lens No. 2, where it would be diverged even more and would be flung out of the system. However, the development so far neglects space-charge forces which are all important to us. The space-charge forces act so as to reduce the convergence of the trajectories between lenses. It is then possible to choose conditions so that the paths between lenses are symmetrical about the planes $z=nL\pm L/2$ and are represented by part of the universal beam spread curve. The procedure is given in detail by Pierce[(8)]. Mendel *et al.*[(30)] consider the behaviour of a magnetic field which is assumed to have a constant value B_0 over a gap length l, separated by L from the next gap where the field is $-B_0$, in detail and they find that the condition $CL<4$ reduces on the insertion of the value of C for a short magnetic lens, i.e.

$$C=\eta/8V_0\int_0^z B_z^2\,.\,dz,$$

the integral being taken along the axis of the tube, to

$$\eta\,\frac{B_1^2L^2}{V_0}<64, \tag{89}$$

where B_1 is the amplitude of the fundamental of the axial induction and is equal to

$$\frac{4B_0(2l/L)}{I_0(4\pi a/L)}\cdot\frac{\sin(2\pi l/L)}{(2\pi l/L)}, \tag{90}$$

I_0 is the Bessel function with imaginary argument, and a is the diameter at which the sudden field changes occur. When space charge is considered, these authors find that the numeric in (89) must be increased to 104.5.

It is more productive, however, to consider the type of focusing from a more general point of view. It is well known that the paraxial ray equation for electrons in *combined* electric and magnetic fields is [see, for example, Beck[10], Chap. 5, eq. (32)]

$$\frac{d^2r}{dz^2}+\frac{V_0'}{2V_0}\frac{dr}{dz}+\frac{1}{4V_0}\left(V_0''+\eta\frac{B_z^2}{2}\right)r=0. \tag{91}$$

We are now interested in finding solutions of eq. (91) valid when the electric field terms or the magnetic field term or combinations of both vary co-sinusoidally along the axis. For example, let us assume that $B_z^2=B_0^2\cos 2z$. We then put

$$\frac{V_0'}{2V_0}=2\kappa,\quad \frac{V_0''}{4V_0}=\bar{a}\quad\text{and}\quad \eta\frac{B_0^2}{8V_0}=-2q,$$

so that (91) can be rewritten

$$r''+2\kappa r'+(\bar{a}-2q\cos 2z)r=0.$$

Then put

$$y=e^{-\kappa z}.r \tag{92}$$

to obtain

$$y''+(a-2q\cos 2z)\,y=0 \tag{93}$$

where

$$a=\bar{a}-\kappa^2.$$

Equation (93) is Mathieu's equation in the standard form extensively discussed by MacLachlan[32]. The solution of eq. (93) is

$$y=\exp(\mu z).\phi(z) \tag{94}$$

or

$$r=\exp(\mu-\kappa)z.\phi(z). \tag{95}$$

When $\mu-\kappa$ is positive, eq. (95) shows that the trajectories increase without limit in the z direction, i.e. are unstable, but when $\mu-\kappa$ is negative, periodic solutions are obtained, which are those of interest in electron optics. The reader should consult MacLachlan for further details of the solutions, which have been fairly extensively tabulated.

When space charge is important, we have to subtract the additional term $(I_0/4\sqrt{2\varepsilon_0}\eta^{1/2}V^{3/2})(1/r)$ from the left-hand side of (91). The equation will not then transform into a standard Mathieu equation but it can still be computed fairly easily. However, by the time we have reached this stage we are concerned with matters which are still the subject of detailed development.

13. ANNULAR BEAMS

Annular electron beams are at present the subject of intensive research and if good and efficient means of launching such beams become available, they will assume great practical importance. The reasons for their importance are:

1. For a given tunnel diameter a thin annular beam will carry much more current for a specified potential depression than will a beam filling the whole tunnel. The physical reason for this is that the charge is concentrated near to the conductor and is therefore less effective in depressing the potential than is an equal charge on the axis of the tube. The ratio of currents depends on the dimensions of the annulus, but it can easily be of the order 10.

2. In microwave valves it is found that current on the axis of a beam is far less effective than is current on the periphery. The detailed reasons for this are complicated and need not detain us here, but it is a fact that annular beams may be useful even when their outer diameter is several times as large as the limiting diameter for a solid beam.

14. THEORY

We now consider an annular electron beam of circular cylindrical shape constrained by a very strong magnetic field. The beam moves inside a hollow conductor at potential V_0. We treat explicitly the case in which there is no centre conductor as this is the one usually met in microwave valves, where the centre conductor would complicate the r.f. field configuration. The notation is shown in Fig. 17. If the charge density inside the beam is assumed to be uniform, then the expressions for the potential are the same as those developed earlier for a magnetically confined beam, that is, the potential inside the beam is given by eq. (39), i.e.

$$\frac{d^2\Phi_1}{dR^2}+\frac{1}{R}\frac{d\Phi_1}{dR}=\Phi_1^{-1/2}. \qquad (96)$$

We redefine R for this case by

$$R = \frac{r}{(b^2-c^2)^{1/2}} K^{1/2}\left(\frac{V_0}{V_m}\right)^{3/4}. \tag{97}$$

For the region $b<r<a$, the potential is given by

$$\frac{d^2\Phi_2}{dR^2}+\frac{1}{R}\frac{d\Phi_2}{dR}=0, \tag{98}$$

while inside the beam $r<c$, Gauss's theorem tells us that the potential is constant and that $E_r=0$ at $r=c$. Thus the potential V_m is now the potential anywhere inside the beam, which is better written as V_c.

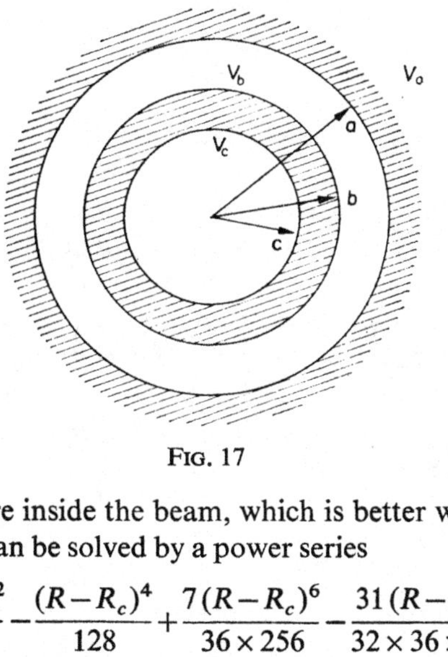

FIG. 17

Equation (96) can be solved by a power series

$$\Phi_1 = 1+\frac{(R-R_c)^2}{4}-\frac{(R-R_c)^4}{128}+\frac{7(R-R_c)^6}{36\times 256}-\frac{31(R-R_c)^8}{32\times 36\times 256}, \text{ etc.,} \tag{99}$$

where $R_c = R$, evaluated at $r=c$.

We next have to make the potential and its derivatives with respect to R continuous at $r=b$. Either by integrating (98) subject to the boundary conditions or by the use of Gauss's theorem, the potential in region (2) is found to be

$$V = V_0+\frac{(V_b-V_0)\ln a/r}{\ln a/b} \tag{100}$$

or

$$\Phi_2 = \Phi_0+\frac{(\Phi_b-\Phi_0)\ln a/r}{\ln a/b}. \tag{101}$$

Thus

$$\left[\Phi_2'\right]_b = \frac{\Phi_0 - \Phi_b}{R_b \ln a/b} = \left[\Phi_1'\right]_b \qquad (102)$$

together with

$$[\Phi_1]_b = 1 + \frac{(R_b - R_c)^2}{4} - \frac{(R_b - R_c)^4}{128}, \text{ etc.} \qquad (103)$$

Since

$$R_b = \frac{b}{\sqrt{(b^2 - c^2)}} K^{1/2} \left(\frac{V_0}{V_c}\right)^{3/4}$$

we can rewrite eq. (102)

$$\left[\Phi_1'\right]_b = \left[\frac{V_c}{V_0}\right]^{3/4} \frac{1}{K^{1/2}} \left(\frac{b^2}{b^2 - c^2}\right)^{1/2} \frac{1}{\ln a/b} \left[\frac{V_0}{V_c} - [\Phi_1]_b\right]. \qquad (104)$$

The procedure for handling these equations is as follows. Φ_1 and Φ_1' are computed as functions of R either by digital computation or by the use of the power series, eq. (99), for sufficiently small values of R. Values of a/b and b/c are then chosen and eq. (104) is solved for the values of $K^{1/2}$ giving predetermined values of V_0/V_c. Extensive graphs of these results are given by Brewer[(33)]. His results show that the potential at the inner edge of the beam falls monotonically as the perveance increases, until the maximum perveance is reached when the inner potential is 25–30 per cent of the wall potential and the beam becomes unstable. Examples of his curves are shown in Fig. 18 a and b. However, we may note that the most useful beams are those in which the drop in potential across the beam is small, of the order of 10–20 per cent, so that Φ does not much exceed unity. $(R_b - R_c)$ does not then exceed unity and the power series is useful.

As an example, suppose we wish to find the perveance which will make the wall potential 110 per cent of the inner potential when the beam just grazes the outer tube, i.e. $b \to a$. Then $\Phi = 1.1$ and from eq. (99), $(R_b - R_c)^2 \doteqdot 0.4$. But from the definitions of R_b and R_c, $(R_b - R_c)^2 = [(b-c)/(b+c)]K(V_0/V_c)^{3/2}$, and $K = 6.04 \times 10^4 p$, where p is the perveance expressed with respect to the wall potential.

Then

$$p\left(\frac{b-c}{b+c}\right) = \frac{40 \times 10^{-6}}{6.04 \times (1.1)^{3/2}} = 6.28 \times 10^{-6}$$

so that if $c = 0.8b$, $p = 56.5 \times 10^{-6}$.

We have seen that for the solid beam filling the tunnel, the maximum perveance is only 31.5×10^{-6}, while for a 10 per cent potential variation this figure is reduced to 6×10^{-6}. The annular beam is therefore capable of carrying nine times the current of the solid beam, thus substantiating our introductory remarks.

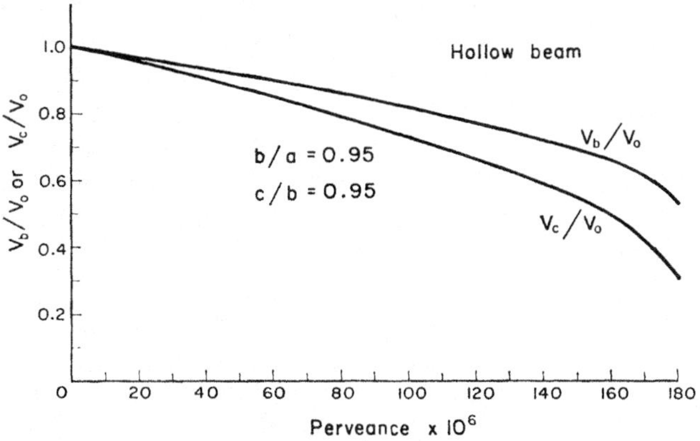

(a)

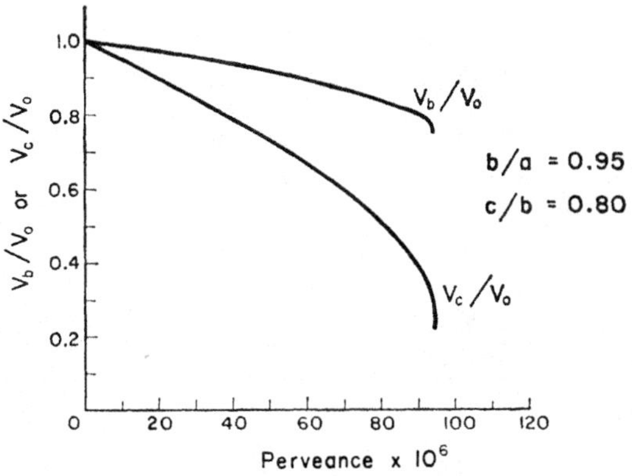

(a)

FIG. 18

The advantages of magnetically confined annular beams are now clear. The reason that such beams have not been extensively used lies in the gun. Many attempts have been made to design Pierce-type guns which would focus the electrons originating on a thin annular cathode of mean diameter large compared with that of the beam, into a thin annular beam. To the best of my knowledge all attempts to do this have met with so little success that they have been abandoned. The reason for this lack of success is probably that it is a very difficult problem to realize electrostatic fields which will satisfy all the following requirements simultaneously. Firstly, the fields must be strong enough to draw off the required current. Secondly, they must direct electrons on the inner and outer edges of the beam along trajectories which are not only different but are not even similar. Thirdly, the conditions at entry to the final magnetic field must be correct.

Practical annular beam guns as used until recently have therefore been based on the use of a plane cathode facing a grid with an annular aperture making no attempt either to decrease the mean diameter of the beam, or to thin it down. For some purposes, especially for low frequency valves, in which the mean diameter of the annulus is large, such guns are effective, but they are too restricted to find general application. Some work has been done by the author and his colleagues(34) and by other laboratories(35) on methods of reducing the mean diameter of an annular beam formed in this way and it has been demonstrated that this can be successfully done, but finality has by no means been reached.

15. MAGNETRON INJECTION SYSTEM

An alternative method for producing annular beams has long been dormant in the literature. In 1945 Brillouin(21) described a type of flow obtained from a unipotential cathode inside a cylindrical anode (Fig. 19). In recent years several laboratories have investigated this

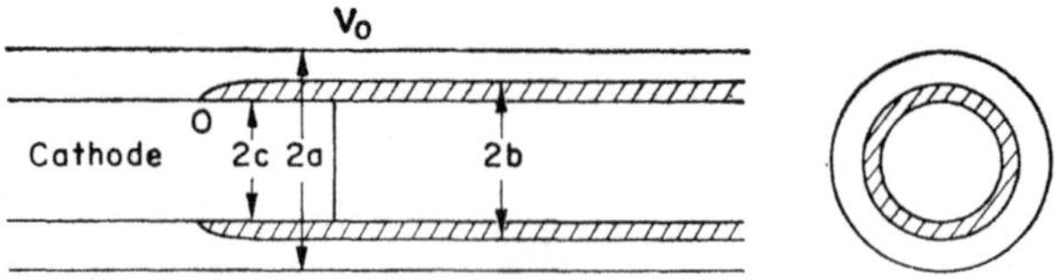

FIG. 19

system experimentally, among them the Hughes Aircraft Laboratory (Todd and Brewer(36)) and the Engineering Laboratories of Cambridge University (Beck and Nattrass(36a). These studies show that the magnetron gun is a very useful method for producing high perveance beams, $6\text{–}30 \times 10^{-6}$ quoted by Todd and Brewer, while the guns of Beck and Nattrass gave $p = 20\text{–}80 \times 10^{-6}$ and perveances per unit square of about unity. On the other hand, the detailed behaviour of the annular beams produced is not at all in agreement with the Brillouin theory and much work on the detailed elucidation of the trajectories remains to be done.† However, the Brillouin theory serves as a useful starting point and we shall therefore outline it here.

In his 1945 paper, Brillouin[(21)] shows, starting from the correct Larmor form of the Lagrangian function for the electron, that various conservation principles could be proved. Clearly, the total energy is conserved as magnetic fields do no work, but there are also conservation principles applicable to momentum and to angular momentum.

For example, consider the situation of Fig. 19 where we show the cross-section of a magnetron in which the anode and cathode diameters, a and d respectively, are of similar magnitude. Consider an electron at a point x_0, y_0 on the cathode surface and neglect thermal velocities.

The Lagrangian is

$$L(x, y, z, \dot{x}, \dot{y}, \dot{z}, t) = 1/2mu^2 - e\phi + e(u \,.\, A), \tag{105}$$

with

$$u \,.\, A = \dot{x}A_x + \dot{y}A_y + zA_z. \tag{106}$$

The momentum has components given by

$$p_x = \frac{\partial L}{\partial \dot{x}}, \text{ etc.}$$

If now we set up our magnetron so that $B_x = 0 = B_y$ and $B_z = B_0$, a constant, we can show, from $\boldsymbol{B} = \text{curl}\,\boldsymbol{A}$, that $A_z = 0$, $A_x = -(B_0/2)y$, $A_y = (B_0/2)x$.

The angular momentum at the point of origin on the cathode is

$$x_0 p_y - y_0 p_x = x_0 \left[\frac{\partial L}{\partial \dot{y}}\right]_{y_0} - y_0 \left[\frac{\partial L}{\partial \dot{x}}\right]_{x_0}.$$

Carrying out the partial differentiations we have

$$m\dot{y}_0 x_0 - m\dot{x}_0 y_0 + \frac{eB_0}{2} x_0^2 + \frac{eB_0 y_0^2}{2}$$

† For recent work see refs. (66) and (67).

and since $\dot{x}_0 = 0 = \dot{y}_0$ and $x_0^2 + y_0^2 = d^2$ the angular momentum becomes

$$\frac{eB_0}{2m} d^2 = -\omega_L d^2. \tag{107}$$

Here we have introduced the Larmor angular frequency $\omega_L = \eta B_0/2$ as usual. Since angular momentum is conserved in this system, it must always have the constant value $\omega_L . d^2$.

We can now find the angular velocity as a function of radius from

$$r^2\dot{\theta} = \omega_L(r^2 - d^2),$$

i.e.

$$\dot{\theta} = \omega_L\left[1 - \left(\frac{d}{r}\right)^2\right]. \tag{108}$$

Finally, we inquire if we may have a stable electron flow with finite $\dot{\theta}$ and zero $\dot{r}$, so that the trajectories are circles round the cathode. Since energy is conserved and $\dot{r} = 0$, the energy relation is

$$2\eta V_r = r^2\dot{\theta}^2 + u_z^2 = \frac{\omega_L^2}{r^2}(r^2 - b^2)^2 + u_z^2,$$

therefore

$$u_z^2 = 2\eta V_r - \frac{\omega_L^2}{r^2}(r^2 - b^2)^2 \tag{109}$$

or, for the anode ($r = a$),

$$u_z^2 = 2\eta V_r - \frac{\omega_L}{a^2}(a^2 - b^2)^2. \tag{110}$$

Thus the flow is possible if $(\omega_L/a^2)(a^2 - b^2)^2 < 2\eta V_a$. But $2\eta V_a = (\omega_L/a^2)(a^2 - b^2)^2$ is the ordinary Hull cut-off condition, so we must conclude that the flow yields finite values of u_z when the magnetic field is below the Hull cut-off value and that $u_z = 0$ at cut-off.

In fact, the work on magnetron injection shows that there is a z-directed electron flow when the magnetic field is above the cut-off value and since this is the condition in which very few or no electrons are lost to other electrodes, this is a desired operating region. Thus we have to seek an explanation of the flow observed for $B > B_c$.

However, for the moment let us avoid this question and instead consider the current, magnetic field voltage relation which exists for Brillouin flow in an annular beam.

16. BRILLOUIN FLOW IN AN ANNULAR BEAM

We first, for simplicity, consider the case when the annular beam just grazes the inside of the surrounding tunnel so that the outer beam radius $b =$ tunnel radius a. We omit any discussion of the injection system but require that the annular beam satisfies the Brillouin conditions, viz. that the projection of the trajectories on the (x, y) or (y, z) planes are straight lines, i.e. that the electrons on the inner radius $r = c$ travel parallel with the axis while the electrons at $r = b$ perform helical paths of constant diameter round the axis. As a consequence, we shall find that the z-directed velocity is the same for all electrons.

For a balance of radial forces we must have

$$eE_r + eB_B r\dot{\theta} - mr\dot{\theta}^2 = 0. \tag{111}$$

At the inner radius $r = c$, $E_r = 0$ from Gauss's theorem, so that

$$\text{either} \quad \dot{\theta}_c = 0 \quad \text{or} \quad \dot{\theta}_c = \eta B_B. \tag{112}$$

But, from eq. (108) we see that, if the electrons have their origin within the region filled by B_B, they cannot have an angular velocity greater than ω_L, i.e. greater than $\eta B_B/z$. Therefore, the correct solution is $\dot{\theta}_c = 0$.

Therefore

$$\dot{\theta} = \omega_L \left\{1 - \left(\frac{c}{r}\right)^2\right\}. \tag{113}$$

The energy equation thus becomes

$$U = 2\eta V = r^2\dot{\theta}^2 + u_z^2 = \omega_L^2 r^2 \left\{1 - \left(\frac{c}{r}\right)^2\right\}^2 + u_z^2. \tag{114}$$

But

$$-E_r = \frac{1}{2\eta}\frac{\partial U}{\partial r} = \frac{\omega_L^2}{\eta}\left\{r - \frac{c^4}{r^3}\right\} + \frac{u_z}{\eta}\frac{\partial u_z}{\partial r}. \tag{115}$$

However, from eq. (111) we must also put

$$-E_r = \frac{\omega_L^2}{\eta}\left\{r - \frac{c^4}{r^3}\right\} \tag{116}$$

so that $\partial u_z/\partial r = 0$ or u_z is independent of r. Clearly, from the use of eq. (114) at $r = c$, u_z is the velocity equivalent to the potential at the inner radius of the beam, which differs from the tunnel potential by the depression due to the total space charge.

We now use eq. (116) to calculate the space-charge density from Poisson's equation, which in cylindrical polars reduces to

$$\frac{1}{r}\,\partial/\partial r\,(rE_r) = \rho/\varepsilon_0$$

or

$$\rho = \frac{2\omega_L^2\varepsilon_0}{\eta}\left\{1+\frac{c^4}{r^4}\right\}. \tag{117}$$

We are now able to calculate the beam current from

$$I = 2\pi u_z \int_c^a \rho r\,\mathrm{d}r = \frac{2\pi\varepsilon_0\omega_L^2 u_z}{\eta}\left(\frac{a^4-c^4}{a^2}\right). \tag{118}$$

Finally, u_z is found from taking the energy value at $r = a$, or

$$2\eta V_a = \omega_L^2\left[\frac{a^2-c^2}{a}\right]^2 + u_z^2. \tag{119}$$

Since we usually require to calculate the magnetic field required to produce a Brillouin beam using a given current I, it is convenient to write eq. (118) as

$$B_B^2 = \frac{2I}{\eta\pi\varepsilon_0\,.\,u_z}\left(\frac{a^2}{a^4-c^4}\right). \tag{120}$$

We next inquire, using eqs. (119) and (120), what maximum current can be obtained for fixed a, c and V_a with this particular flow. The details are straightforward but the algebra is tedious so we merely quote the results.

The maximum current is obtained when $u_{z(\mathrm{opt})}^2 = 2/3\eta V_a$ or, in words, when the axial potential is one-third of the value when there is no current. The corresponding maximum current is

$$\hat{I} = \frac{4\pi\varepsilon_0}{\eta}\left(\frac{2\eta V_a}{3}\right)^{3/2}\left(\frac{a^2+c^2}{a^2-c^2}\right) \tag{121}$$

or

$$\hat{I} = 25.4\times10^{-6}\,V_a^{3/2}\left(\frac{a^2+c^2}{a^2-c^2}\right) \tag{122}$$

when

$$B_B^2 = \frac{2\hat{I}}{\eta\pi\varepsilon_0 \, . \, u_{z(\text{opt})}}\left(\frac{a^2}{a^4-c^4}\right) \tag{123}$$

or

$$B_B^2 = \frac{12.1\times 10^{-7}\,\hat{I}}{\sqrt{V_a}}\left(\frac{a^2}{a^4-c^4}\right). \tag{124}$$

It will be noted that eqs. (122) and (124) reduce correctly to the solid beam values when $c \to 0$.

It is at once obvious that, for a thin beam for which $a-c$ is small, the maximum perveance is very much greater than for a solid beam. This is in line with our findings for annular beams confined by very strong fields.

When the beam is smaller than the surrounding tunnel the analysis is slightly modified as we have to include the logarithmic variation of potential between the outer edge of the beam at $r=b$ and the conductor at $r=a$. Using Gauss's theorem we obtain

$$V_b = V_a - \frac{I_0}{2\pi\varepsilon_0 \, . \, u_z}\ln(a/b). \tag{125}$$

Carrying out the optimization, we again find $u^2_{z(\text{opt})} = (2\eta V_a/3)$ but $\hat{I}$ is now given by

$$\hat{I} = 25.4\times 10^{-6}\, V_a^{3/2}\,\frac{1}{\{[(b^2-c^2)/(b^2+c^2)]+2\ln(a/b)\}} \tag{126}$$

and

$$B_B^2 = \frac{12.1\times 10^{-7}\,\hat{I}}{\sqrt{V_a}}\left(\frac{b^2}{b^4-c^4}\right). \tag{127}$$

To complete the results, it is useful to give the method of determining u_z for currents below $\hat{I}$. If we put $\beta = u_z/(\sqrt{2\eta V_a})$ it can be shown that

$$\beta^3 - \beta + \frac{2I}{3^{3/2}\hat{I}} = 0 \tag{128}$$

so that if we solve eq. (128) for any value of $I/\hat{I}$ we find the required value of u_z. Since β varies between 1 and $(3)^{-1/2}$ it is natural to approximate by putting $\beta = 1-\delta$ and using two-term binomial expansions to obtain $\delta = I/3^{3/2}\hat{I}$.

Thus for $I/\hat{I} < 0.5$,

$$u_z \risingdotseq \sqrt{(2\eta V_a)}\left(1-\frac{I}{3^{3/2}\hat{I}}\right). \tag{129}$$

In conclusion we emphasize that the necessary conditions for setting up an annular Brillouin beam imply a specific variation of charge density with radius, given by eq. (117). An identical charge distribution is found in Brillouin's[37] single stream theory of the magnetron. This theory is known not to account for many phenomena observed in actual magnetrons, but it forms a starting point for the theory of magnetron guns.

17. ELECTROSTATIC FOCUSING OF HIGH CURRENT BEAMS

We have already seen that the simplest type of electrostatic focusing in which the beam profile is determined purely by the space-charge forces is useless for long, high-perveance beams. However, the alternatives which have so far been discussed involve magnetic fields and therefore either permanent or electromagnets for their production. Even when the design is very carefully adjusted these are heavy and expensive in terms of both money and operating power. Considerable effort has, therefore, been devoted to exploring other approaches. Such approaches fall into three major categories.

1. Strong focusing, i.e. the use of a distribution of electrostatic einzellinsen. We have already discussed the magnetic equivalent of this.
2. Methods of focusing in which the beam is set into rotary motion.
3. Combination of electrostatic and magnetic focusing in which a magnetic field, acting over a short distance, is used to start the beam rotating.

We now discuss these in turn.

18. ELECTROSTATIC STRONG FOCUSING

The basic theory is that already discussed, which is found in an amplified form in Pierce[8]. Other important papers are those of Clogston and Heffner[38], Tien[39,40] and of Siekanowicz and Vaccaro[41].

Electrostatic strong focusing exhibits the fundamental difficulty that a high voltage ratio on the lenses is required to obtain a short focal length. While this is tolerable in microscope optics, in heavy current beams the effect of space charge in depressing the potential inside the beam must be taken into account and the perveance is limited by the formation of a virtual cathode in the low potential section of the lens.

The complete theory is complex and many assumptions and simplifications have to be made, but when this is done it is found that a quasi-cylindrical beam with only small scalloping can be maintained. For the simple systems shown in Fig. 20, the perveance is approximated by

$$p = 61 \times 10^{-6} \left(\frac{V_1}{V_0}\right)^2 \left(\frac{2a}{L}\right)^2 \tag{130}$$

for $V_1 \ll V_0$ and $2a \ll L$.

Here, V_1 is the potential $V_H - V_L$ and V_0 is the average potential

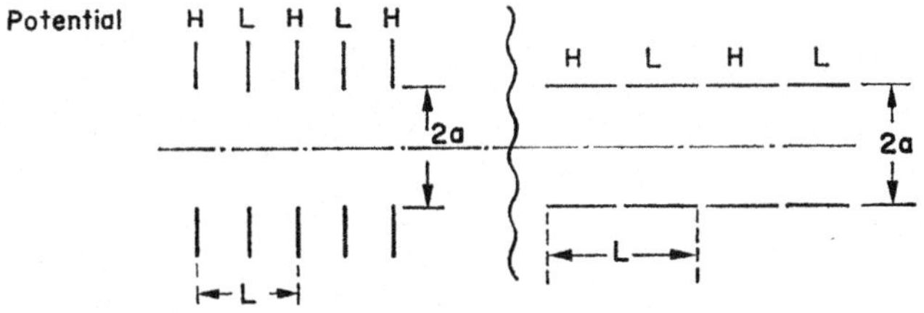

FIG. 20

$(V_H + V_L)/2$. In the derivation, the charge density is assumed uniform as it should be when the beam originates from a well-designed gun. In Tien's work on the use of a bifilar helix with different potentials on the two conductors, the achievement of a quasi-cylindrical beam involves a non-uniform beam density increasing towards the outer edge of the beam. As yet, guns capable of achieving desired density distributions have not been designed and Tien's measured results relate to beams from normal guns. This perhaps accounts for the rather poor transmission reported at a perveance of only 0.1 microperv (78 per cent transmission at 1 mA beam current). Better results have been reported by Blattner, Vaccaro, Cuccia and Johnson[(42)], using considerably higher voltages. For $V_1 \fallingdotseq V_0 = 2$ kV these authors achieved a perveance of 0.56 microperv (beam current = 50 mA) with a transmission efficiency of 99.2 per cent.

Subsequently, Siekanowicz and Vaccaro[(41)] introduced the idea of shaping the focusing electrodes so that each section, starting from a plane of minimum potential, forms a Pierce gun with rectilinear trajectories. Figure 21 illustrates the electrode configuration.

The maximum current turns out to be given by

$$\hat{I} = 14.7 \times 10^{-6} \left(\frac{2a}{L}\right)^2 V_H^{3/2} \tag{131}$$

reached with the condition $V_H = 4.0\ V_L$. An experimental tube with $2a/L = 0.31$, and $V_H = 850\ V$, $V_L = 260\ V$ gave 96 per cent beam transmission for a 30 mA beam. The theoretical perveance was thus 1.47×10^{-6} compared with the experimental value 1.20×10^{-6} which must be considered good agreement.

Considerable ingenuity has to be exercised in designing electrostatic focusing systems for microwave valves because the focusing means have to be compatible with the r.f. structure. For this reason,

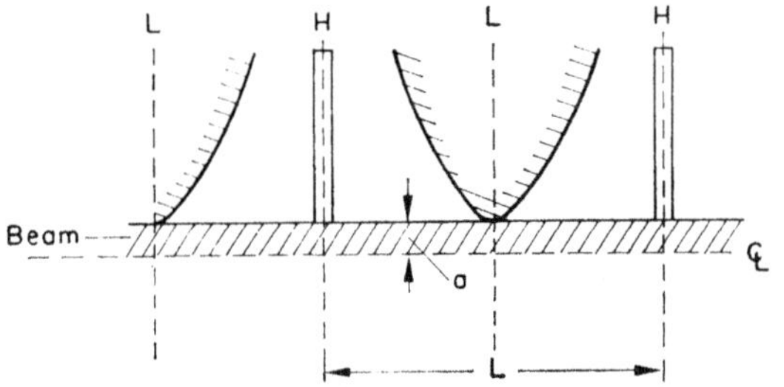

FIG. 21

electrostatic focusing has been little used except for airborne tubes where weight is of major importance. However, it seems likely that there are other applications of electron beams, for example welding and fine machining in which electrostatic systems may have advantages over magnetic focusing. In these high power densities are needed and voltages in the 20–100 kV range are used. Thus, the mean potential is high even when quite large voltage ratios are used. The chief practical difficulty is that of maintaining the insulation on the lenses in the rather poor vacuum conditions encountered in the systems under consideration. Examples of electron optical problems encountered in this field are described in ref. 43.

Recently, the needs of airborne systems have led to the reintroduction of multi-resonator klystrons using electrostatic einzellinsen. Tubes of this type were first described by Touraton, Zwobada and Dumousseau[43a] in 1947. Subsequently, much higher powered tubes of this type have been designed by groups led by Hechtel at Litton Industries, who have made a 100 W output tube in the 2300 Mc band with a bandwidth of 35 Mc/sec, a very good performance, and at Siemens, where a 1 MW *S*-band pulsed tube has been produced. The latter tube is only about one-fifth of the weight of a similar design using magnetic focusing. These tubes are described in the *Proceedings of the Sixth International Conference on Microwave and Optical Generation and Amplification*, Cambridge, 1966.

19. SYSTEMS IN WHICH THE BEAM ROTATES

We have already said something about the system due to Chang[2] in which the beam originates from a gun immersed into a magnetic field and emerges through an aperture in the magnetic circuit into a region free of magnetic fields. This starts the beam rotating and gives

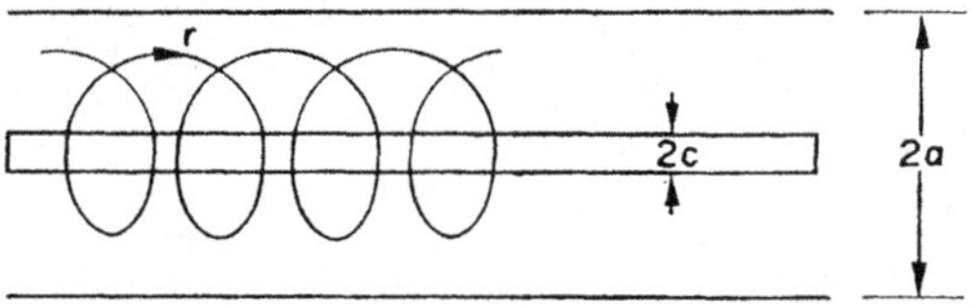

FIG. 22

an extra centrifugal force term, which when inserted into the equations of electrostatic strong focusing permits the attainment of a very stable flow. However, there exist some purely electrostatic systems which allow us to obtain similar results. One basic element is common to several of these. This is the use of the field between coaxial cylindrical conductors held at different potentials to achieve a stable flow with helical trajectories (Fig. 22). When the inner conductor potential is higher than the outer, so that the force eE_r on the electrons is inwards, a balance between centrifugal and electrostatic forces is clearly possible in the single electron case. Correct conditions for the injection can readily be worked out if it is assumed that this can be done from outside the cylinders, for the gun can be arranged for fire in a thin beam tangentially to the helix path and the angle between the axis of the gun and the xy plane can be adjusted to give any desired z-directed velocity. The accelerating potential must be adjusted to provide the sum of the rotational and longitudinal energies. A system of this sort has been used by Watkins and Wada[44] in the Helitron, a simple type of B.W.O., but the maximum current achieved was only 1 mA at 1700 V with an optimum transmission efficiency of only 43 per cent.

Much better results have been achieved by Chernov and his collaborators[45] who launch the beam as shown in Fig. 23. Chernov[46] has shown that the calculated perveance depends on what assumption is made about the processes controlling the instability which sets in as the perveance increases. The form of the perveance is always the same, viz.

$$I_0 = \frac{33 \times 10^{-6} V_f \sqrt{V_z}}{C} \tag{132}$$

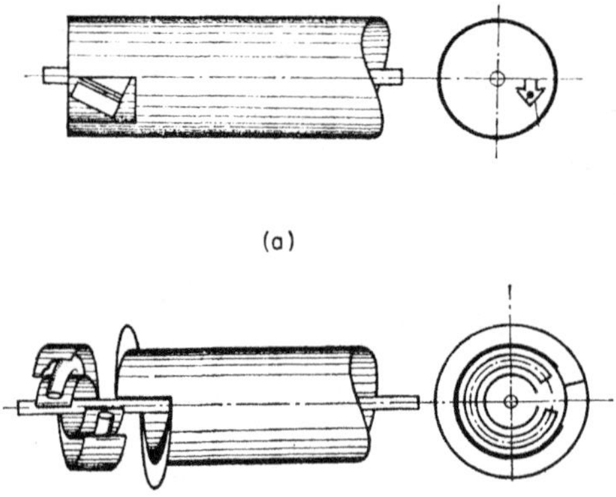

(a)

(b)

FIG. 23

but different values of C are obtained. Three causes of instability are considered: (1) reduction of the electric field at the outer beam edge to zero, (2) scalloping of the outside surface of the beam, (3) when the perturbed trajectories of certain electrons lose stability although the configuration of the beam as a whole is unchanged.

The coefficients are defined by

$$C_1 = \ln(b/d) - 1/2 + k \ln(b/c), \tag{133}$$

$$k = \frac{c^2}{b^2 - a^2},$$

$$C_2 = \ln(b/d) + k/2 - k^2 \ln(b/c), \tag{134}$$

$$C_3 = \ln(b/d) - 1/2 - k\ln(b/c) + \frac{b^2}{b^2 - c^2}\ln(a/d). \qquad (135)$$

Here a is the inner diameter of the outer conductor, b is the outer diameter of the beam, c is the inner diameter of the beam, and d is the outer diameter of the centre conductor. V_f is the p.d. between conductors and V_z the axial velocity in volts.

For a given geometry $1/C_3$ is the *smallest* of these coefficients and experimental results indicate that the corresponding perveance is in fairly good agreement with results observed on beams for which the transmission efficiency was about 90 per cent.

Experimental results showing transmission efficiencies of 98 per cent for currents of 25 mA at potentials of 1.5 and 1.8 kV, i.e. perveance $\risingdotseq 0.4 \times 10^{-6}$, have been quoted.

Another combination of magnetically induced rotation with electrostatic focusing is due to Harris[(47)]. Here, as in Chang's system, the beam is generated in a magnetically shielded enclosure and it is set rotating by the flux cut in emerging from the enclosure. Instead of maintaining the beam by periodic focusing, a two-potential coaxial cylinder system is used. By this means high perveance beams of up to 16×10^{-6}, with transmission efficiencies of about 60 per cent, were obtained. Harris flow has the disadvantage, compared with Chang flow, of requiring the centre conductor, which introduces complications into the r.f. circuitry and prevents the use of this flow in several well-known types of circuit.

20. STRIP BEAMS

Strip beams have found little application outside the field of devices using crossed electric and magnetic fields, but here they are invariably used. In view of the specialized nature of this application we shall only comment on the more important features of crossed field guns, referring the reader to the literature for more details. Early crossed field electron guns were based on designs arrived at by considering trajectories of single electrons, i.e. neglecting space charge. They took the general form of Fig. 24, refs. 48, 49. The plate is operated at a voltage of about $V_0/2 - V_0/3$ where V_0 = circuit potential. Single trajectory considerations, however, were soon found not to give correct information on the optimum cathode breadth, which is very important from the viewpoint of loading and life. In particular, experiment showed that cathodes could be made several cyclotron wavelengths long, whereas the single trajectory picture would indicate that at one cyclotron wavelength electrons emitted at the back should hit at the front, as in the dotted trajectory. Earlier, Brillouin had suggested a space-charge balanced flow which could exist in a linear magnetron[50] and Stainsby[51] has considered crossed field

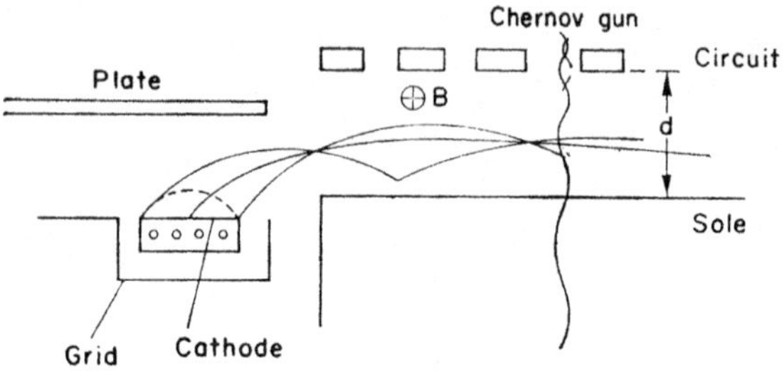

FIG. 24

guns from this point of view. The Brillouin flow has the following properties:

1. Space-charge density is constant.
2. Flow is linear.
3. Velocity and electric field vary linearly across the beam from sole to beam boundary. Potential variation is thus square law.
4. Cathode field is zero, i.e. boundary conditions are as for Child's law.

If the height of the interaction space is d, Stainsby defines the current voltage characteristic in terms of I_{max}, the current just filling the interaction space at the cut-off induction B_c. B_c is related to voltage by

$$V_0 = \eta \frac{B_c^2 d^2}{2} \tag{136}$$

and

$$I_{max} = 1/2\, \varepsilon_0 \eta^2 B_c^3 d^2. \tag{137}$$

Then, for any lower voltage V, it can be proved that

$$V/V_c = \left(\frac{I}{I_{max}}\right)^{1/2}\left[2-\left(\frac{I}{I_{max}}\right)^{1/2}\right]. \tag{138}$$

Equation (138) can easily be tested experimentally and is found to agree very closely with the characteristics of actual guns. It is therefore concluded that the experimental flows are of this general type.

Ideas of the same general character have led Kino(52) to extend the Brillouin trajectories back into the gun region, thus producing

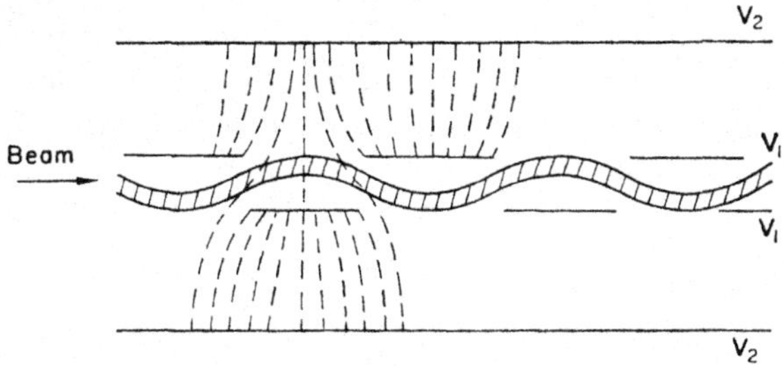

FIG. 25

an analytical method for designing guns of this type. Midford and Kino(53) have described experiments on guns designed in this way.

Strip beam guns have occasionally been developed for use in magnetically focused klystrons. An example with rather good performance is described by Cutting and Fraser(54). This gun gives a current of 40 mA at a potential of 1.0 kV which converges so that the transmission efficiency through a rectangular tunnel 0.4 mm wide,

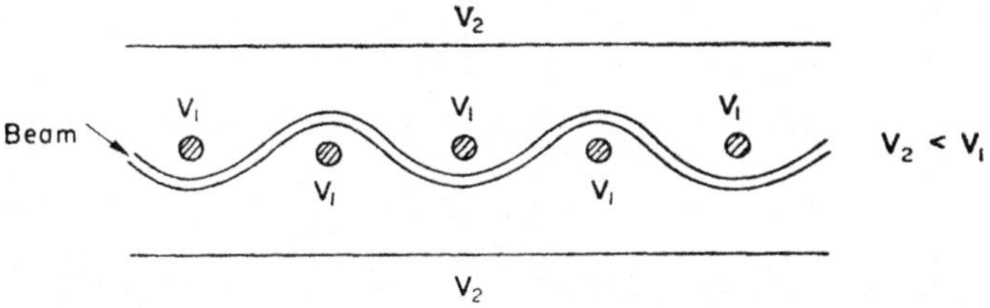

FIG. 26

3.0 mm high and 7.0 mm long is about 85 per cent. The induction was 0.2 Wb/m.

Strip beams can be maintained by purely electrostatic means; for instance, by passing the beam through an array of electrodes at alternating high and low potentials. The electrodes on either side of the beam can be in-line or offset. A very ingenious modification is due to Hogg(55). This, as shown in Fig. 25, allows the electrodes of the ladder delay line all to be at the high potential V_1 while only the focusing structure V_2 is at a lower voltage. A slightly modified version of this focusing, using only a single ladder, has been used in a B.W.O. called the "Ophitron". Finally, Kompfner[56] and other workers at the Bell Laboratories[57,58] have described "slalom" flow, which can be achieved using the structure of Fig. 26.

In general it appears that the obvious advantage of strip beams, that the total current can be increased without limit by increasing the width, is outweighed by the difficulties of incorporating such beams in apparatus and perhaps by difficulties in achieving correct end conditions in the gun.

21. NON-LAMINARITY

All the theory so far used has been based on the assumption that electron beams exhibit laminar flow, i.e. that electron trajectories do not cross inside the beam. Although it has been well known for at least 20 years that at least some guns produced beams which were far from laminar, no systematic work was done on this question until Harker[59] studied the phenomenon in solid beams emanating from magnetically shielded cathodes in circumstances where Brillouin flow would be expected or, at least, hoped for. By traversing a pinhole along the beam, Harker demonstrated that ring images were formed when the pinhole was at appropriate axial positions. Thus electrons were crossing the axis at these points. Harker did not put forward an explanation for the non-laminarity which was so clearly demonstrated by this experiment, but other workers at B.T.L. including Hermann[28] and Ashkin[27] extended this work. Hermannn has described the optical theory of beam formation for guns with magnetically shielded cathodes in which it is assumed that the electron trajectories in the accelerating region between cathode and first anode depend strongly on the initial emission velocity and direction. Thus, at a plane where simple theory would predict a "crossover", there is actually a disc of least confusion with a Gaussian current distribution. Subsequently, in the region of uniform magnetic and zero electric field, a succession of cathode images are formed. The optical theory is able to predict:

(1) that the minimum focusing field will be greater than the minimum Brillouin value;
(2) that the ripple or scallop wavelength is different from the Brillouin value;
(3) that results obtained by Harker are reasonable.

However, optical theory has some difficulties, one of the more important of which is the fact that in many high perveance electron guns, especially those of low convergence, it is readily proved by trajectory tracing that thermal velocities have very little influence on the trajectories. Also, magnetic fields in the cathode region very much diminish the effects of thermal velocities, as in fact is mentioned by Hermann. At present, we must therefore conclude that the optical theory may describe some low perveance high convergence guns rather well but its application to other systems needs justification.

22. SOME EFFECTS OF THERMAL VELOCITIES

We must now take up the question of the effects of thermal velocities. Until recently it was imagined, on energetic considerations, that thermal velocity spread produced no important effects in electron beams which had been accelerated through more than about 300 V. It is now known that if the convergence is high and the perveance is low, a situation encountered in guns for millimetric tubes, thermal velocities may produce readily measurable effects even at voltages of 1–2 kV.

The influence of low perveance is readily explicable, since perveance is intimately connected with transit time in the anode–cathode space. To make this statement obvious, consider a space-charge limited diode in the Child's law approximation. The transmit time τ is 50 per cent greater than in the same diode when it is temperature limited (Beck, ref. 10, p. 186).

Therefore

$$\tau = 3d/u_0. \tag{139}$$

Child's law can then be written in the form (Σ = cathode area)

$$I_0 = J_0\Sigma = \frac{2\varepsilon_0 u_0 \Sigma}{\tau^2 \eta}. \tag{140}$$

But

$$I_0 = pV_0^{3/2}, \qquad u_0 = 2\eta V_0,$$

therefore

$$\tau^2 = 2\sqrt{(2/\eta)}\,\frac{\varepsilon_0 \Sigma}{pV_0} \tag{141}$$

or
$$\tau = \kappa \sqrt{\left(\frac{\Sigma}{pV_0}\right)}. \tag{142}$$

Next, we observe that the transverse velocity due to thermal energy is conserved during the transit so that the transverse deflexion δ is given by

$$\delta = u_y . \tau$$
$$= \kappa u_y . \sqrt{\left(\frac{\Sigma}{pV_0}\right)}. \tag{143}$$

Thus the deflexion due to thermal velocities varies directly as the square root of the cathode area and inversely as the square root of (pV_0). This product varies widely in current electron optical devices. For example, in electron microscopes $p = 10^{-10}$–10^{-11}, $V_0 = 10^4$–10^6 so $(pV_0)^{1/2}$ varies from 10^{-2} to 3×10^{-3}, while in klystrons $p = 10^{-6}$–3×10^{-6} and V_0 varies from 10^3 to 4×10^5 so that $(pV_0)^{1/2}$ ranges from 3×10^{-2} to 1.1. Of course, the guns used depart very much from planar diodes, but the calculation does show that thermal effects are likely to be important in the low current devices even at very high voltages unless Σ is very small. This is probably the reason why large cathode Pierce guns have been relatively unsuccessful when applied to cathode ray tubes.

Cutler and Hines[(60)] have applied these ideas to the calculation of the effects of thermal velocities in Pierce guns. The relation between radial coordinate and time is now transcendental involving Langmuir's analysis and, in particular, functions of $(-\alpha^2)$. The results therefore depend on detailed numerical calculations and are not susceptible to simple expression as are those of the planar diode, but the results lead to the same general conclusion—the throat calculated after making allowance for thermal velocities is considerably greater in diameter and is nearer to the gun than is the throat calculated purely on a space-charge basis, even for quite high voltages.

The effect of convergence is also readily stated, although a calculation has to appeal to more advanced ideas. In the field free region beyond the anode aperture the flow is quasi-laminar, i.e. laminar complicated by stray electrons. Since total energy has to be conserved, it is obvious that the greater the range of angles with which electrons traverse an infinitesimal area, the greater the range of transverse velocities. Cutler and Hines[60] are able to show that if the cathode density obeys the Maxwellian law

$$dJ_c = A\,e^{-(m/2kT)(u_x^2+u_y^2)}\,du_x\,du_y \qquad (144)$$

then the density at plane z is

$$dJ_z = A\,e^{-(m/2kT)M_z^2(u_x^2+u_y^2)}\,du_x\,du_y, \qquad (145)$$

where M_z = beam radius at z/beam radius at cathode. We may thus say that the transverse electron temperature has been increased by $1/M^2$ and if the convergence is large ($M_z \ll 1$), this can be a large amount. We may comment in passing that Liouville's theorem, which states that the density of particles adjacent to a trajectory in phase space remains constant, immediately leads to the general conclusion that a reduction in the coordinate range d^3q must be accompanied by an increase in the momentum (velocity) range d^3p in phase space.

The work of Cutler and Hines has been extended by Danielson, Rosenfeld and Saloom[61], who have produced tables which show the influence of thermal velocities on spherical cap guns of a range of perveances. These authors also consider the effect of the anode aperture in reducing the gun perveance, but only rather low perveance guns are treated. Further extensions have more recently been made by Kirstein[62].

CONCLUSION

This article has been mainly devoted to the problems encountered when one wishes to choose an electron gun suitable for some particular experiment. It has not attempted to go into the detail of, for example, computer solutions either analytical or synthetic. These are highly specialized questions which seemed best omitted from a handbook article. Some extremely precise work on electron/ion guns has also been done in the field of ionic propulsion of rockets[63], which we lack space to include. It is hoped that the information given will make it easier for the non-specialist to utilize better electron guns and to avoid some well-known pitfalls.

REFERENCES

1. PIERCE, *J. Appl. Phys.* **10**, 715 (1939).
2. CHANG, *Proc. I.R.E.* **45**, 1522 (1957).
3. PIERCE, *J. Appl. Phys.* **11**, 548 (1940).
4. LANGMUIR and BLODGETT, Concentric cylinders, *Phys. Rev.* **22**, 347 (1923); Concentric spheres, *Phys. Rev.* **24**, 49 (1924).
5. BECK, JACKSON and PREDDY, Report B.R.S. 508/45/XQ2 (1945).
6. BARBER and SANDER, *J. Electronics and Control*, **7**, 465 (1959).
7. BECK and DEERING, *Proc. I.E.E.* **105 B**, 833 (1958).
8. PIERCE, *Theory and Design of Electron Beams*, 2nd ed., p. 174 *et seq.*, Van Nostrand, 1954.
9. BOWMAN-MANIFOLD and NICOLL, *Nature*, **139**, 1066 (1937).
10. BECK, *Thermionic Valves*, Chap. IV, Cambridge U.P., 1953.
11. SPANGENBERG, HELM and FIELD, *Elec. Commun.* **24**, 108 (1947).
12. DAVISSON and CALBICK, *Phys. Rev.* **38**, 585 (1931); **42**, 580 (1932).
13. SAMUEL, *Proc. I.R.E.* **33**, 233 (1945).
14. HUBER, *Annls Radioélectr.* **4**, 26 (1949).
15. WATSON, *Phil. Mag.* **3**, 849 (1927).
16. THOMPSON and HEADRICK, *Proc. I.R.E.* **28**, 318 (1940).
17. MILLER and GORDON, *J. Chem. Phys.* **3**, 285 (1935).
18. SMITH and HARTMANN, *J. Appl. Phys.* **11**, 220 (1940).
19. GUÉNARD, *J. Phys. Radium*, **6**, 260 (1945).
20. BREWER, Hughes Aircraft Co. ETL Memo. 55–26 (May 1955).
21. BRILLOUIN, *Phys. Rev.* **67**, 260 (1945).
22. BECK, *J. Electronics*, **2**, 489 (1957).
23. MÜLLER, *Telefunken Zeits.* **26**, 95 (1953).
24. RIGROD, *B.S.T.J.* **36**, 831 (1957).
25. HARKER, *J. Appl. Phys.* **28**, 645 (1957).
26. TUDOR JOHNSON, *Proc. I.E.E.* **105 B**, 907 (1958).
27. ASHKIN, *J. Appl. Phys.* **29**, 1594 (1958).
28. HERMANN, *J. Appl. Phys.* **29**, 127 (1958).
29. KLEEN and PÖSCHL, *A.E.U.* **9**, 295 (1955).
30. MENDEL, QUATE and YOCOM, *Proc. I.R.E.* **42**, 800 (1954).
31. KELLY, Mullard Research Report No. 2367 (Jan. 1962).
32. MACLACHLAN, *Theory and Applications of Mathieu Functions*, Oxford U.P., 1947.
33. BREWER, Hughes Aircraft Co. ETL Memo. 55–26 (May 1955).
34. DEERING, S.T.L. Memo. Unpublished.
35. DUNN and HOLLADAY, *Proc. 3rd International Congress on Microwave Valves*, p. 452, Munich, 1960.
36. TODD and BREWER, *I.R.E. Conference on Electron Tube Research*, Mexico City, 1959.

36a. BECK and NATTRASS, Microwaves, *Proc. 4th International Conference on Microwave Valves*, p. 599, Central Pub. Co., Eindhoven, 1962.

37. BRILLOUIN, *Elec. Commun.* **20**, 112 (1941); *Phys. Rev.* **60**, 385 (1941); *Proc. I.R.E.* **32**, 216 (1944).
38. CLOGSTON and HEFFNER, *J. Appl. Phys.* **25**, 436 (1954).
39. TIEN, *J. Appl. Phys.* **25**, 1281 (1954).
40. TIEN, *Proc. I.R.E.* **42**, 1137 (1954).

41. Siekanowicz and Vaccaro, *Proc. I.R.E.* **47**, 451 (1959).
42. Blattner, Vaccaro, Cuccia and Johnson, *R.C.A. Rev.* **20**, 4626 (1959).
43. *Proc. 3rd Symposium on Electron Beam Technology*, 1961, Alloyd Electronics Corp., Boston, Mass., 1961.
43a. Touraton, Zwobada and Dumousseau, *Onde Électrique*, **27**, 361 (1947).
44. Watkins and Wada, *Proc. I.R.E.* **46**, 1700 (1958).
45. Chernov, *Proc. I.E.E.* **105** B, 594 (1958).
46. Chernov, Tubes hyperfréquences, *Travaux du Congres International*, Vol. II, p. 27, Paris, 1956.
47. Harris, *Proc. I.R.E.* **40**, 700 (1952).
48. Warnecke, *Proc. I.R.E.* **38**, 486 (1950).
49. Davis and Sander, *J. Electronics and Control*, **5**, 114 (1958).
50. Brillouin, *Adv. Electronics*, **3**, 85 (1952).
51. Stainsby, *Proceedings 3rd International Conference on Microwave Valves*, p. 489, Munich, 1960.
52. Kino, *I.R.E., P.G.E.D.*, ED-**7**, 179 (1960).
53. Midford and Kino, *I.R.E., P.G.E.D.*, ED-**8**, 324 (1961).
54. Cutting and Fraser, Tubes hyperfréquences, *Travaux du Congres International*, Vol. II, p. 191, Paris, 1956.
55. Hogg, *Proc. I.E.E.* **105** B, 1016–20 (1958).
56. Cook, Kompfner and Yocom, Tubes hyperfréquences, *Travaux du Congres International*, Vol. II, p. 36, Paris, 1956.
57. Cook, Kompfner and Yocom, *Proc. I.R.E.* **45**, 1517 (1957).
58. Cook, Louisell and Yocom, *J. Appl. Phys.* **29**, 583 (1958).
59. Harker, *J. Appl. Phys.* **28**, 645 (1957).
60. Cutler and Hines, *Proc. I.R.E.* **43**, 307 (1955).
61. Danielson, Rosenfeld and Saloom, *B.S.T.J.* **35**, 375 (1956).
62. Kirstein, Dubna Reprint E 1023 (1962).
63. *Progress in Astronautics and Rocketry*, Vol. 5 (1961), Vol. 9 (1963), Academic Press, N.Y.

ADDITIONAL REFERENCES

64. Brewer, High density beams, *J. Appl. Phys.* **28**, 7 (1957).
65. Frost, Pierce and Johnson, High perveance guns of large convergence ratio, *Proc. I.R.E.* **50**, 1800 (1962).
66. Tubes pour hyperfréquences, *Proceedings 5th International Conference on Microwave Valves*, Dunod, Paris, 1964.
67. *Proceedings 6th International Conference on Microwave and Optical Generation and Amplification, Cambridge, 1966.* I.E.E., London, 1967.

PART 6

GRIDDED VALVES. THEORY AND DESIGN PRINCIPLES

A. H. BECK

Engineering Laboratory, Cambridge

PART 6

GRIDDED VALVES. THEORY AND DESIGN PRINCIPLES

A. H. BECK

Engineering Laboratory, Cambridge

1. GENERAL ASSUMPTIONS

In valve design several assumptions are *always* made and it is convenient to enumerate these now. First, we assume that the vacuum is sufficiently high for us to be able to neglect effects due to the presence of gas ions, unless these are specifically required and means for their production are provided. This assumption means that the electronic mean free path (m.f.p.) in the apparatus under discussion must be much greater than the longest path taken by the electrons moving through the device. It must naturally be remembered that magnetic fields may increase the path length by large factors so that our requirement is much more stringent when magnetic fields are used than it is in their absence.

As a rule of thumb, it is worth remembering that m.f.p. of the molecules in air at 20°C is 7 cm at pressure of 1 μ. The electronic m.f.p. is 5.6 times as great, i.e. about 40 cm. Mean free paths vary inversely as the pressure, so that at 10^{-6} torr the electronic m.f.p. is about 400 m.

Ions accumulate if no sweeping fields are provided to remove them and in discussing long electron beams it seems to be necessary to use pressures below 10^{-5} torr if first order effects due to ions are to be negligible.

Our second assumption is that the electron velocities are low enough for the relativistic increase in mass to be negligible. An approximation for the relativistic velocity, valid for small β ($\beta = u_0/c$) is

$$u_0 = \sqrt{(2\eta V_0)}\left(1-\frac{3\beta^2}{8}\right). \qquad (1)$$

Comparing this with the classical velocity

$$u_0 = (2\eta V_0), \tag{2}$$

we find that the true velocity is about 7 per cent less than the classical value for 50 kV electrons. Some beam devices, especially electron microscopes, work at higher voltages and for these the mass increase may have to be considered. Although it will be unimportant in the sequel, we may emphasize that consistency requires us also to neglect the self-magnetic field due to the current flow in an electron beam.

We also very often assume that secondary emission effects may be neglected. Usually, the effects of secondary emission are deleterious, the obvious exception being in secondary emission multipliers and in secondary emission storage devices. Thus, considerable pains are often taken to eliminate the secondary electrons which inevitably appear whenever an electron beam is collected at a final anode or elsewhere. Efficient collection may be very difficult or even impossible in practice, but we assume that, if necessary, means can be included to remove secondaries.

2. AN OUTLINE OF VALVE THEORY

In the theory of conventional valves in which the current through the valve is controlled by the potential applied to one or more electron-permeable electrodes, called grids, we should like to be able to use purely analytical solutions of Poisson's equation

$$\nabla^2 V = -\rho/\varepsilon_0 \tag{3}$$

which would reduce to the prescribed values of potential on the various electrodes. Except in one particular case, that of the planar diode, and even then only with simplifying assumptions, this programme is much too ambitious for present-day mathematics and we have to proceed by breaking the problem into two parts. We first determine the potential distribution in the valve, i.e. we solve Laplace's equation

$$\nabla^2 V = 0 \tag{4}$$

subject to the prescribed potentials on the boundaries and we then calculate an electron flow which will be consistent with the fields which we have determined.

Laplace's equation can, of course, be solved by a number of techniques, some of which have been discussed at other places in this book. For example, we can find analytical solutions for some, rather simple, geometries. Analytical solutions are possible in closed form; for example, for planes, concentric cylinders and spheres. In series form, analytical solutions are available for many other arrangements, but in the general case and in the particular case of the gridded valve exact analytical results have been little used. For particular cases, the potential problem is readily solved using an analogue computer; for example, the rubber sheet model or the electrolytic tank(1). More recently, digital computers(2) have been used to calculate the potential distribution using, for example, the relaxation technique of Southwell(3). However, the classical method of valve design is an approximate one in which the electrodes of the real valve are represented as line and cylinder charges by the technique of conformal transformation. This technique, unfortunately, becomes inaccurate when the electrostatic μ of the valve becomes large and, in particular, the technique is useless for modern microwave triodes for which μ ranges from 100 to 400. The best calculations for high μ planar triodes have been made by Bennett and Peterson(4) using an infinite series of images, an idea first suggested by Noether(5). We shall, in fact, later see that the work of Bennett and Peterson suggests that some rather simple laws exist for the design of planar triodes of maximum μ and that further improvements in this field are purely a question of improving manufacturing techniques and materials.

To carry this discussion further we shall discuss, in some detail, the design of planar triodes and show how the principles can be extended to multi-electrode valves. We begin with the simplest case of all, the planar diode.

3. CHILD'S LAW FOR THE PLANAR DIODE

We consider that the cathode and anode planes are of infinite extent so that edge effects may be entirely neglected. The calculation made here was originally given by Child(6), whose contribution was to define the boundary conditions at the cathode in a particular way. His assumptions were that at the cathode surface both the potential V and the field $E_z = -\mathrm{d}V/\mathrm{d}z$ were zero. We can, of course, always take the potential as zero on the cathode. The second condition merits closer attention: we know, from the theory of thermionic emission, that the emitted electrons all possess initial velocities greater than zero, so that if the field at the cathode is finite and is in such a direction as to repel electrons, only a certain fraction of the saturation emission will in fact be observed. On the other hand, if $(dV/dz)_c = 0$, then the saturation emission must flow. (We may, if we like, consider this as a definition of saturation emission.) Now, it is very unlikely that the current through our diode will happen to be the saturation current from the cathode and if it does so happen, saturation can only occur at a particular voltage. Thus Child's

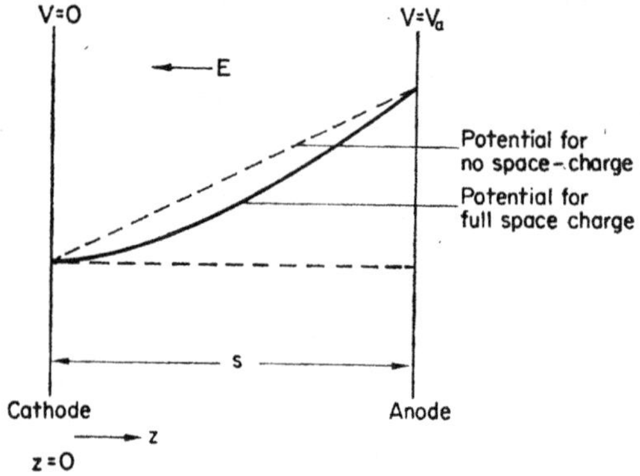

FIG. 1

assumption is fundamentally unsound and cannot be correct except for very special cases. However, experiment shows that the assumption leads to results which are numerically quite close to the truth, except for very closely spaced, low voltage diodes, and it is commonly used to obtain results which, though inexact, are of considerable practical value.

The notation for the problem is given in Fig. 1. In one-dimensional flows, Poisson's equation is

$$\frac{d^2V}{dz^2} = -\rho/\varepsilon_0.$$

Let J be the current density, u be the electron velocity. Then $\rho = -J/u$. But $u = \sqrt{(2\eta V)}$.

Therefore
$$\frac{d^2V}{dz^2} = \frac{J}{\varepsilon_0}\sqrt{\left(\frac{m}{2e}\right)}.V^{-1/2}. \tag{5}$$

Integrating twice we obtain, since from the initial conditions both constants of integration are zero,

$$\frac{4V^{3/4}}{3} = \sqrt{\left[\frac{4J}{\varepsilon_0}\sqrt{\left(\frac{m}{2e}\right)}\right]} . z$$

or

$$J = \frac{4\varepsilon_9}{9\sqrt{(m/2e)}} \cdot \frac{V^{3/2}}{z^2}. \tag{6}$$

Now, eq. (6) holds anywhere along the flow and in particular it must hold at the anode, where $V = V_a$ and $z = s$. Thus

$$= \frac{4\varepsilon_0}{9\sqrt{(m/2e)}} \cdot \frac{V_a^{3/2}}{s^2} \tag{7}$$

$$= 2.336 \times 10^{-6} \frac{V_a^{3/2}}{s^2} \text{ A/m}^2.$$

To obtain the total current $I_c = I_a$ we must multiply by the cathode–anode area.

It is often useful to know how the potential varies with distance along the trajectories. From (7) we find

$$V = 5.68 \times 10^3 J^{2/3} z^{4/3} \tag{8}$$

and since $E = -\mathrm{d}V/\mathrm{d}z$,

$$E \fallingdotseq -7.6 \times 10^3 J^{2/3} z^{1/2}. \tag{9}$$

The electron velocity varies as $\sqrt{V}$, i.e. as $J^{1/3}z^{2/3}$ and the charge density as J/U or as $J^{2/3}z^{-2/3}$.

This completes the derivation of the information which is to be

TABLE 1

s	V_a
0.1 mm	21.5
1.0 mm	465
1.0 cm	10,000
1.0 m	4.65×10^6

obtained from Child's law [eq. (7)]. Let us now consider the magnitudes involved. Most useful thermionic cathodes give saturation emissions of several amps per square centimetre. Let us assume $J_{sat} = 2.336 \times 10^4$ A/m^2. Then, from eg. (7), $V_a^{3/2} = 10^{10}s^2$ giving the results of Table 1.

So that, for spacings greater than about 1 mm very high voltages are required to draw the saturation emission, or, on the other hand, at normal voltages the cathode current will be much lower than the saturation emission. How, then, is the success of Child's law to be explained? The answer is that if the emission velocities are distributed according to a Maxwellian law, the current density due to the number of electrons which will be able to surmount a potential barrier V_m (a potential of $-V_m$ with respect to cathode) is given by

$$J = J_{sat} \exp\left(\frac{-V_m}{kT}\right)$$

$$= J_{sat} \exp\left(-\frac{11{,}606\ V_m}{T}\right), \tag{10}$$

where T is the cathode temperature corresponding with J_{sat}. This lies in the range 800–2500°K so that the exponential argument is never below 4.5 V_m in magnitude. Therefore, for $V_m = 1$ V, the current reaching the anode is never greater than 0.0111 times the saturation emission, even for a bright tungsten emitter. For oxide cathodes, values of V_m around -0.5 V will reduce the anode current to about one thousandth of the emission current. From Table 1 we have seen that quite large values of V_a are required, according to Child's law, to draw the saturation emission. We have thus shown that unless the anode–cathode spacing is very small, the magnitude of the potential minimum cannot exceed a few per cent of the total voltage difference across the diode.

We may note at this point that the whole process of current control in a space-charge control valve entirely depends on the depth of the potential minimum and varying the current must be done by influencing V_m. Also, the electrons which actually reach the anode are those whose thermal energy, inside the cathode, is greater than $e(\phi_c + V_m)$ eV and no other electrons can reach the anode.

From these considerations we see that the thermal energies of the electrons play a vital role in the detailed physics of valves even though simple theories ignoring thermal effects may give results which are sufficiently accurate numerically for engineering work.

Next, we must discuss the theory of the diode with initial velocities taken into account. In the form most commonly used today, this is due to Langmuir[(6)]. Here, we shall give a very brief discussion of the results and shall refer to other sources (refs. 6, 7, 8) for the details.

Figure 2 shows the situation in a valve with a potential minimum of depth V_m at a distance z_m from the cathode. To the right of z_m, all the electrons are travelling towards and will finally reach the anode. To the left of z_m, the flow has two components, first all the emitted electrons moving to the right, secondly all the electrons which possessed energies between ϕ_c and $\phi_c + V_m$ inside the cathode surface, which have been turned round in the region between the cathode and

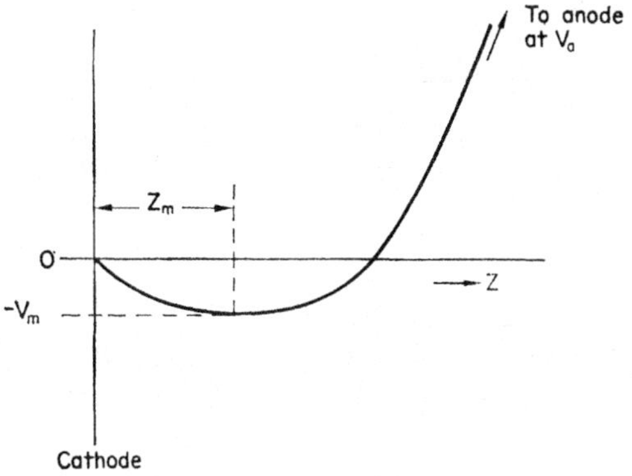

FIG. 2

the plane of the minimum. Poisson's equation can be integrated for the right-hand space to give

$$\left(\frac{dV}{dz}\right)^2 - \left(\frac{dV}{dz}\right)^2_{z=z_m} = -\frac{2}{\varepsilon_0}\int_{V_m}^{V} \rho(z)\,dz \tag{11}$$

and at the minimum the boundary conditions are $V = (-V_m)$ and $E = 0$. But these are the boundary conditions which are used in deriving Child's law, so that we see that the diode with initial velocities behaves very much like a Child's diode in which the voltage is increased to $(V_a + V_m)$ and the spacing is reduced to $(s - z_m)$. Thus an improved version of Child's law is

$$J_a = \frac{2.336 \times 10^{-6}(V_a + V_m)^{3/2}}{(s - z_m)^2}, \tag{12}$$

where V_m must be consistent with eq. (10). We have as yet no information on z_m, but Langmuir's solution allows us to calculate a *maximum* value for z_m. This is

$$z_m = \frac{2.554\,T^{3/4}}{9.186\times 10^5\,J_a^{1/2}}\ \text{m} \qquad (13)$$

z_m, of course, decreases as J_a increases and is by definition zero when the saturation emission density is reached.

TABLE 2. (PART 1). TABLE OF η AGAINST $-\xi$

η	$-\xi$	η	$-\xi$
0.0	0.0	1.1	1.656
0.001	0.0629	1.2	1.708
0.002	0.0887	1.3	1.756
0.003	0.1084	1.4	1.801
0.004	0.1250		
0.005	0.1395	1.5	1.842
0.006	0.1527	1.6	1.881
0.007	0.1647	1.7	1.917
0.008	0.1759	1.8	1.952
0.009	0.1863	2.0	2.013
0.010	0.1926	2.2	2.068
0.011	0.2056	2.4	2.117
0.012	0.2146	2.6	2.160
0.013	0.2231	2.8	2.199
0.014	0.2314	3	2.234
0.015	0.2393	3.2	2.265
0.02	0.2753	3.4	2.293
0.03	0.3351	3.6	2.318
0.04	0.3848	3.8	2.341
0.05	0.4281	4	2.362
0.06	0.4668	4.5	2.404
0.07	0.5019	5	2.438
0.08	0.5346	5.5	2.463
0.09	0.5655	6	2.483
0.1	0.5941	6.5	2.499
0.2	0.817	7	2.511
0.3	0.979	8	2.528
0.4	1.108	9	2.538
0.5	1.217	10	2.544
0.6	1.312	11	2.548
0.7	1.396	12	2.550
0.8	1.470	15	2.553
0.9	1.538	20	2.554
1.0	1.600	∞	2.554

TABLE 2. (PART 2). TABLE OF ξ AGAINST η

ξ	η	ξ	η
0	0	3.7	2.182
0.01	0.000025	3.8	2.282
0.1	0.00245	3.9	2.384
0.2	0.00964	4.0	2.488
0.25	0.01493	4.2	2.700
0.3	0.02131	4.4	2.915
0.32	0.02413	4.6	3.140
0.34	0.02719	4.8	3.370
0.36	0.03047	5.0	3.604
0.38	0.03388	5.2	3.844
0.4	0.03737	5.4	4.092
0.42	0.04100	5.6	4.343
0.44	0.04482	6.0	4.856
0.46	0.04880	6.5	5.526
0.48	0.05299	7.0	6.224
0.5	0.05726	7.5	6.948
0.55	0.0689	8.0	7.697
0.6	0.0813	8.5	8.471
0.65	0.0946	9.0	9.272
0.7	0.1089	9.5	10.09
0.75	0.1239	10	10.93
0.8	0.1401	11	12.67
0.85	0.1571	12	14.49
0.9	0.1748	13	16.38
0.95	0.1934	14	18.32
1.0	0.2127	15	20.35
1.1	0.2538	16	22.43
1.2	0.2980	17	24.60
1.3	0.3451	18	26.76
1.4	0.3954	19	29.03
1.5	0.4481	20	31.35
1.6	0.5036	21	33.71
1.7	0.5617	22	36.09
1.8	0.6223	23	38.53
1.9	0.6854	24	41.04
2.0	0.7510	25	43.56
2.1	0.8190	26	46.16
2.2	0.8891	27	48.77
2.3	0.9612	28	51.41
2.4	1.036	29	54.14
2.5	1.112	30	56.86
2.6	1.191	32	62.42
2.7	1.272	34	68.09
2.8	1.354	36	74.02
2.9	1.439	40	86.11
3.0	1.526	50	118.6
3.1	1.614	60	153.5
3.2	1.704	80	230.2
3.3	1.796	100	313.8
3.4	1.890	120	403.8
3.5	1.986	140	499.7
3.6	2.083	160	600.9
3.7	2.182		

An improved approximation is

$$J_a = \frac{2.336\times 10^{-6}(V_a+V_m)^{3/2}}{(s-z_m)^2}\left\{1+\frac{2.66}{\sqrt{\eta}}\right\}, \tag{14}$$

where†

$$\eta = \left(\frac{V_a+V_m}{kT}\right) = \frac{11{,}606}{T}(V_a+V_m). \tag{15}$$

Here, the term in parentheses expresses the increase in current due to the increase in initial velocities. If we use eq. (10) in eq. (14), we obtain

$$J_s \exp\left(-\frac{V_m}{kT}\right) = \frac{2.336\times 10^{-6}(V_a+V_m)^{3/2}}{(s-z_m)^2}\left\{1+\frac{2.66}{\sqrt{\eta}}\right\}. \tag{16}$$

When the saturation current flows, V_m and Z_m are both zero and $\eta = 11{,}606 V_a/T$.

Therefore

$$J_s = \frac{2.336\times 10^{-6}\, V_a^{3/2}}{s^2}\left\{1+2.66\sqrt{\left(\frac{T}{11{,}606\, V_a}\right)}\right\}. \tag{17}$$

If we take a diode of known spacing s, the right-hand side of (17) can be plotted and the saturation current can be determined as the intersection of the $J_a \sim V_a$ characteristic of the real valve with the theoretical characteristic [eq. (17)]. The characteristic of the real valve will always lie above the theoretical characteristic when the anode current is below the saturation value.

The design of diodes, using Langmuir's theory, is done using accurate tables of the solutions to the complete Poisson equation. An excellent set of tables is that in refs. 8 and 9. The technique of using the tables is simple if the anode current is known and the spacing or voltage are to be determined. The tables, partially reproduced in Table 2, give η, as defined in eq. (15), as a function of distance from cathode defined in terms of the parameter ξ where

$$\xi = 2L(z-z_m) \tag{18}$$

and

$$L = \frac{(m\pi)^{1/4}}{(2kT)^{3/4}}\left(\frac{eJ}{\varepsilon_0}\right)^{1/2} = 9.186\times 10^5\, T^{-3/4} J^{1/2}\, \mathrm{m}^{-1}. \tag{19}$$

† Following Langmuir, η is *not* here e/m.

The design procedure is then as follows:

1. Calculate V_m from eq. (10).
2. Calculate η at the cathode, i.e. $\eta_c = 11{,}606 V_m/T$.
3. Use the tables to find the value of ξ_c corresponding with η_c.
4. Find $2L$ from eq. (19) for the known T and J.
5. Find $(z - z_m)$ from eq. (18) using the fact that $z = 0$ at the cathode or $\xi_c = 2Lz_m$.
6. Repeat steps (2) and (3) for the anode and calculate the spacing from $\xi_c + \xi_a = 2Ls$.

When the anode current is not known, it is best to use the approximate result [eq. (12)] and to use the tables to refine this.

We notice that, as stated in the preamble to this section, Langmuir's solution only differs numerically from Child's law when V_a and s are both very small, a case which is only encountered in rectifiers for u.h.f. use.

4. CYLINDRICAL DIODES

For diodes with cylindrical symmetry a closed analytical solution cannot be obtained even for the Child's initial conditions. For these Langmuir finds

$$J_c = \frac{2.336 \times 10^{-6}\, V_a^{3/2}}{r_c \cdot r_a \cdot \beta^2} \text{ A/m}^2 \text{ cathode area;} \tag{20}$$

$$\beta = \ln(r/r_c) - \frac{2\ln^2(r/r_c)}{5} + \frac{11\ln^3(r/r_c)}{120} - \frac{47\ln^4(r/r_c)}{3300}, \text{ etc.,} \tag{21}$$

and is given in Table 3. Cylindrical diodes are never used with spacings sufficiently small to necessitate the use of more accurate results.

5. PLANAR TRIODES

We shall work out the design procedure for a planar triode in fair detail, merely quoting the results for cylindrical structures. This procedure is not as arbitrary as it seems because modern valves tend to approximate more and more exactly to the planar form. The procedure is first to solve the potential problem as discussed earlier. This solution besides giving the potential distribution also gives the electrostatic amplification factor $\mu = -\left|\partial V_a/\partial V_g\right|I_{a=\text{const.}}$ of the valve, i.e.

TABLE 3. LANGMUIR'S β^2 FOR CYLINDRICAL DIODES.
β^2 applies where $r > r_c$, $-\beta^2$ applies where $r < r_c$,

r/r_c or r_c/r	β^2	$-\beta^2$	r/r_c or r_c/r	β^2	$-\beta^2$
1.00	0.0000	0.0000	3.8	0.6420	5.3795
1.01	0.00010	0.00010	4.0	0.6671	6.0601
1.02	0.00039	0.00040	4.2	0.6902	6.7705
1.04	0.00149	0.00159	4.4	0.7115	7.5096
1.06	0.00324	0.00356	4.6	0.7313	8.2763
1.08	0.00557	0.00630	4.8	0.7496	9.0696
1.10	0.00842	0.00980	5.0	0.7666	9.887
1.15	0.01747	0.02186	5.2	0.7825	10.733
1.20	0.02815	0.03849	5.4	0.7973	11.601
1.30	0.05589	0.08504	5.6	0.8111	12.493
1.40	0.08672	0.14856	5.8	0.8241	13.407
1.50	0.11934	0.2282	6.0	0.8362	14.343
1.60	0.1525	0.3233	6.5	0.8635	16.777
1.70	0.1854	0.4332	7.0	0.8870	19.337
1.80	0.2177	0.5572	7.5	0.9074	22.015
1.90	0.2491	0.6947	8.0	0.9253	24.805
2.0	0.2793	0.8454	8.5	0.9410	27.701
2.1	0.3083	1.0086	9.0	0.9548	30.698
2.2	0.3361	1.1840	9.5	0.9672	33.791
2.3	0.3626	1.3712	10.0	0.9782	36.976
2.4	0.3879	1.5697	12.0	1.0122	50.559
2.5	0.4121	1.7792	16.0	1.0513	81.203
2.6	0.4351	1.9995	20.0	1.0715	115.64
2.7	0.4571	2.2301	40.0	1.0946	327.01
2.8	0.4780	2.4708	80.0	1.0845	867.11
2.9	0.4980	2.7214	100.0	1.0782	1174.9
3.0	0.5170	2.9814	200.0	1.0562	2946.1
3.2	0.5526	3.5293	500.0	1.0307	9502.2
3.4	0.5851	4.1126	∞	1.000	∞
3.6	0.6148	4.7298			

it gives the value of μ appropriate when the density of free charges is vanishingly small. When the potential problem has been solved, the current flow problem is solved by introducing an equivalent diode, that is, a diode in which the field at the cathode surface and the derivative of this field with respect to the cathode potential are both equal to the values of the same quantities evaluated in the triode. In symbols,

$$E_c \text{ (diode)} = E_c \text{ (triode)}, \tag{22}$$

$$\frac{dE_c}{dV_c} \text{(diode)} = \frac{dE_c}{dV_c} \text{(triode)}. \tag{23}$$

Once the equivalent diode has been determined it is a simple matter of partial differentiation to find the conductance and anode slope resistance of the equivalent diode. These quantities are then equated to the mutual conductance and anode slope resistance of the triode to which the diode considered is equivalent. This argument presents certain difficulties; for example, should allowance be made for the optical transparency, etc.? To discuss such questions is beyond our present purpose and we take the viewpoint that the theory as developed below is sufficient (a) to show the main behaviour of actual valves and (b) to give initial designs which can be improved by judicious experiment.

6. THE ELECTROSTATIC PROBLEM

As a starting point we have to determine the electrostatic fields, which are solutions of Laplace's equation, and to determine the μ of the valve. Nowadays, using analogue or digital computers, it would be quite practicable to determine the field configurations in large numbers of structures and to investigate the μ's. However, traditionally one investigates the solution by analytical means and, for the structures used in valves, only two analytical methods have been of much value. These are (a) the use of conformal transformations to reduce a complicated structure in one complex plane to a simpler one in another, and (b) the use of image theory to replace the electrodes by infinite arrays of image charges, the resulting potential problem being soluble by elliptic functions.

The second approach, in the hands of Bennett and Peterson[(4)], has yielded the most significant results on very high μ triodes known to the author, but it is more complex than is required for a knowledge of the general behaviour of triodes and we consider it no further.

The first approach, which originated with Maxwell's studies of electrostatic screening, shows the main features of valve behaviour without undue mathematical complexity and we shall discuss this in more detail. The first approximation to the behaviour of a real triode is obtained by applying the conformal transformation

$$W = \text{const.} \ln (Z) \tag{24}$$

to a planar triode as shown in *F*. If the constant is taken as $a/2\pi$, then the region width a, centred on a particular grid wire, transforms into the interior of a cylinder, the cathode contracting to a point at the centre of the cylinder and the grid wire to a circular cylinder with centre at (1, 0) (Fig. 3). These statements are proved as follows.

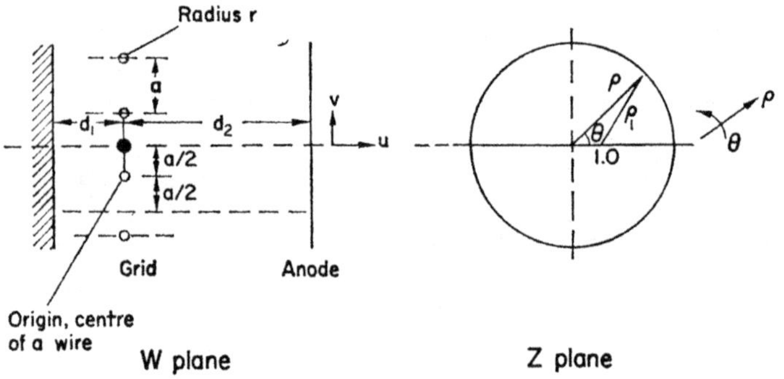

FIG. 3

From eq. (24)

$$u + jv = a/2\pi \ln (\rho\, e^{j\theta})$$

Therefore

$$u = a/2\pi \ln \rho, \quad v = \frac{a\theta}{2\pi}$$

or

$$\rho = e^{2\pi u/a}, \qquad \theta = \frac{2\pi v}{a}. \tag{25}$$

At the grid wire centre, $u = 0 = v$, so that $\rho = 1$, $\theta = 0$ as stated. At the cathode, $u = -d$, so $\rho_c = e^{-2\pi d_1/a}$. In all good valves $d_1 \gg 0.8a$ and therefore ρ_c is very small. At the anode $u = d_2$ and v runs from $-a/2$ to $a/2$. Therefore $\rho_a = e^{2\pi d_2/a}$ (very large) and θ varies between $-\pi$ and $+\pi$.

We can thus represent the cathode by a line charge at the origin and the grid by a line charge at (1, 0), which in turn allows us to write down the potential at an arbitrary point inside the valve in terms of these charges and a constant of integration, which has to be evaluated by making the potential take on the value V_a when the arbitrary point is on the anode. Taking $V_c = 0$ then allows us to solve for the (unknown) cathode and grid charges. This in turn allows us to find μ, using the fact that the cathode charge must be zero if $(V_a + \mu V_g) = 0$. The result is

$$\mu = \frac{-2\pi d_1}{a \ln (2 \sin \pi r/a)}. \tag{26}$$

This result is only valid for very small values of μ. A much better

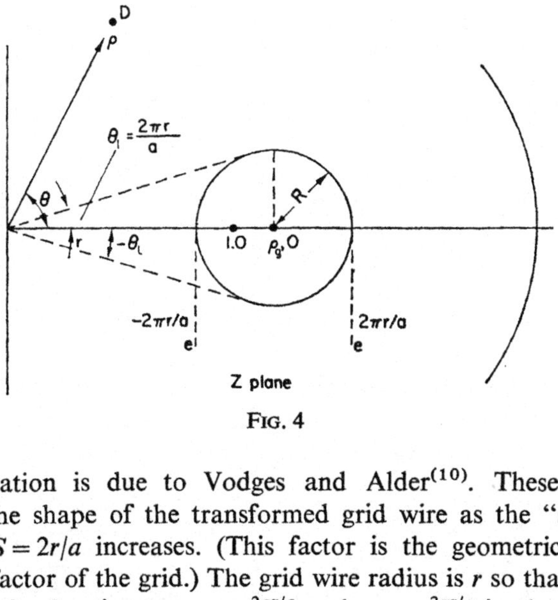

FIG. 4

approximation is due to Vodges and Alder[10]. These authors studied the shape of the transformed grid wire as the "screening factor" $S = 2r/a$ increases. (This factor is the geometrical transparency factor of the grid.) The grid wire radius is r so that it intersects the $\theta = 0$ axis at $\rho_1 = e^{-2\pi r/a}$ and $\rho_2 = e^{2\pi r/a}$ in the Z plane. In the θ direction it extends from $v = -r$ to $v = +r$, i.e. $\theta = \pm 2\pi r/a$. When $2\pi r/a$ is small enough the exponentials can be approximated by $1 + 2\pi r/a$ and the wire can be represented by a circle centred at (1, 0) as before (Fig. 4). As S increases, the grid wire still remains circular but the centre moves out to

$$\rho_g = 1/2\,(e^{\pi S} - e^{-\pi S}) = \sinh \pi S \tag{27}$$

while the radius increases to

$$R = 1/2\,(e^{\pi S} + e^{-\pi S}) = \cosh \pi S. \tag{28}$$

These results are valid for $S > 0.16$. For larger values of S the transformed grid wire assumes a cardioid shape and the analysis breaks down. Dow[(11)] handles the problem of the circular but displaced grid wire by replacing the conductors by three line charges, the cathode charge q_k, the image charge in the grid due to the cathode charge, and the image in the grid due to the anode charge, assumed to be at the centre of the transformed grid wire. Actually, the image of the anode charge is at the *cathode* (the Z plane origin) but the anode radius is so large that shifting this charge to the grid centre

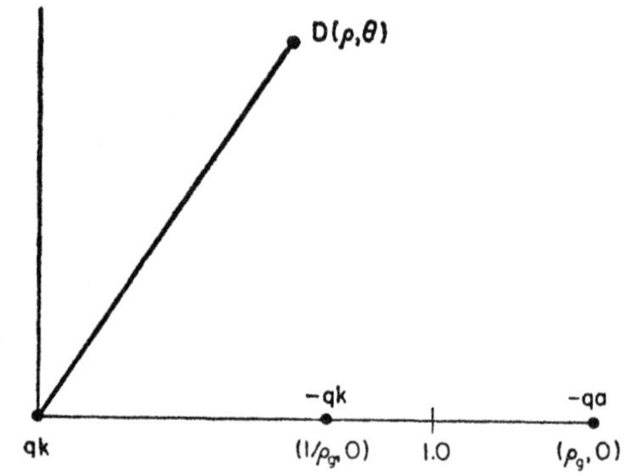

FIG. 5

does not alter the numerical results. From the image theory of a line charge outside a cylinder, the cathode image $-q_k$ is at

$$\rho' = \frac{\rho_g^2 - R^2}{\rho_g}. \tag{29}$$

Using eqs. (27) and (28), $\rho' = 1/\rho_g$.

The potential at a point D, coordinates (ρ, θ) is then (Fig. 5)

$$V_D = \frac{1}{4\pi\varepsilon_0}\left[q_k \ln \rho^2 - q_k \ln\left(\rho^2 + \frac{1}{\rho_g^2} - \frac{2\rho}{\rho_g}\cos\theta\right) - q_a \ln(\rho^2 + \rho_g^2 - 2\rho\rho_g \cos\theta) + C \right]. \tag{30}$$

Let the transformed cathode be a circle radius ρ_k, then on the cathode $V_k = 0$ so that eq. (30) $= 0$ for $\rho = \rho_k$, $\theta = 0$.

Therefore

$$C = -q_k \ln \rho_k^2 + q_k \ln\left(\rho_k - \frac{1}{\rho_g}\right)^2 + q_a \ln (\rho_k - \rho_g)^2. \tag{31}$$

Next, we use the grid and anode potentials to determine the charges. The grid potential is V_g and we have $\rho = \rho_g - R$, $\rho_g^2 = 1 + R^2$, $\rho_k \ll \rho_g$ or $1/\rho_g$. Using these together with eqs. (31) and (30) we find

$$V_g = \frac{1}{2\pi\varepsilon_0}[q_k \ln (R\rho_k) + q_a \ln (R/\rho_g)]. \tag{32}$$

At the anode, $V = V_a$, $\rho = \rho_a$, $\rho_k \ll 1$, $\rho_a \gg 1$, so that

$$V_a = \frac{1}{2\pi\varepsilon_0}[q_k \ln (\rho_k \rho_g) + q_a \ln (\rho_a/\rho_g)]. \tag{33}$$

These last two equations can be solved for q_k and q_a, obtaining

$$q_k = 2\pi\varepsilon_0 \left[\frac{V_g \ln (\rho_a/\rho_g) - V_a \ln (R/\rho_g)}{\ln (R\rho_k) \ln (\rho_a/\rho_g) - \ln (R/\rho_g) \ln (\rho_k \rho_g)}\right], \tag{34}$$

$$q_a = 2\pi\varepsilon_0 \left[\frac{V_a \ln (R\rho_k) - V_g \ln (\rho_k \rho_c)}{\ln (R\rho_k) \ln (\rho_a/\rho_g) - \ln (R/\rho_g) \ln (\rho_k \rho_g)}\right]. \tag{35}$$

From eq. (34), to make $q_k = 0$ we must have

$$\mu = -\ln (\rho_a/\rho_g)/\ln (R/\rho_g) \tag{36}$$

$$= \frac{-\ln \cosh \pi S + 2\pi d_2/a}{\ln \coth \pi S}. \tag{37}$$

Equation (37) is the required result giving μ in terms of the W plane geometry. It should be noted that the terms are all normalized with respect to the grid wire spacing a, so that if *all* the linear dimensions are increased by the same factor n, the μ is unchanged. Equation (37) is accurate up to $S = 0.16$, where the grid shape ceases to transform to a circle. For $S = 0.16$ and $d_2/a = 5$ (this value is about the largest one would use in a valve useful at high frequencies), eq. (37) gives $\mu \doteqdot 40$. Other results are shown in Fig. 6. Real microwave triodes have μ's of up to 350.

Physically, eq. (37) is defective in that it is independent of d_1/a, which is an important parameter. A very great amount of effort has been expended on improving μ formulae and a summary of this is given by Beck[(1)]. However, it is now well known that the best results are obtained when the grid–cathode spacing is between 1 and 0.8 times the spacing between grid wires. If this is to be very small, the grid wires must be made very thin, unless the optical transparency of the grid is to be very low. The problem of triode design is therefore

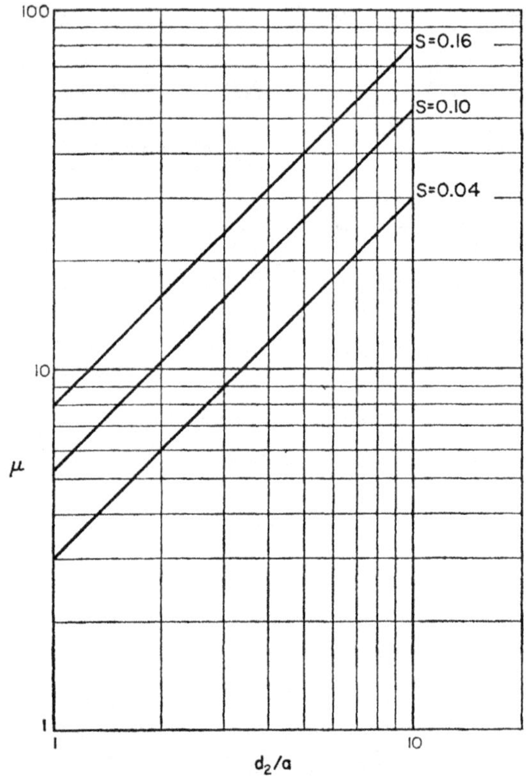

FIG. 6. Graph of μ vs. d_2/A.

reduced to that of mechanical design in drawing, handling and stretching the very thin wires, down to 0.00033 in., now used.

7. THE EQUIVALENT DIODE

We are now able to set up the diode which is equivalent to our triode and from this we shall be able to calculate $g = (\partial I_a/\partial V_g)_{V_a = \text{const.}}$ and $r_a = (\partial V_a/\partial I_a)_{V_g = \text{const.}}$ for the triode.

We must first, however, find the field at the cathode when this electrode is not at zero potential but at some arbitrary value, V_k. Since $E_k = q_k/a\varepsilon_0$ we can generalize eq. (34) to give

$$E_k = \frac{2\pi}{a}\left\{\frac{V_a - V_k + \mu(V_g - V_k)}{\mu \ln(R\rho_k) + \ln(\rho_g \rho_k)}\right\}, \tag{38}$$

therefore

$$\frac{\partial E_k}{\partial V_k} = \frac{-2\pi}{a}\left\{\frac{1+\mu}{\mu \ln(R\rho_k) + \ln(\rho_g \cdot \rho_k)}\right\} \tag{39}$$

while for the diode, which we intend to make equivalent to the triode,

$$E_k = -1/d\,(V_D - V_R), \tag{40}$$

$$\frac{\partial E_k}{\partial V_k} = 1/d. \tag{41}$$

Here, V_D = anode potential of the diode, d = diode spacing.

We now equate (41) and (39) to obtain

$$d = -a/2\pi\left(\frac{\mu \ln(R\rho_k) + \ln(\rho_g \cdot \rho_k)}{1+\mu}\right) \tag{42}$$

or, inserting the transformed dimensions,

$$d = -\frac{a}{2\pi}\frac{1}{1+\mu}\left\{\mu\left(\ln \sinh \pi S - \frac{2\pi d_1}{a}\right) - \frac{2\pi d_1}{a} - \ln \cosh \pi S \right. \tag{43}$$

$$\fallingdotseq d_1\left[1 + \frac{d_2}{(1+\mu)\,d_1}\right]. \tag{44}$$

Equation (44) is obtained as follows: from eq. (37)

$$\mu \ln \coth \pi S = \frac{2\pi d_2}{a} - \ln \cosh \pi S$$

or

$$\mu(\ln \cosh \pi S - \ln \sinh \pi S) = \frac{2\pi d_2}{a} - \ln \cosh \pi S.$$

Thus

$$d = d_1\left[1 + \frac{d_2}{(1+\mu)\,d_1} - \frac{a}{2\pi d_1}\left(\frac{\mu+2}{\mu+1}\right)\ln \cosh \pi S\right].$$

But $\pi S \leqslant 0.6$ so $\cosh \pi S = 1 + (\pi S)^2/2$ and $\ln \cosh \pi S = (\pi S)^2/2$. Since $a/d_1 < 1.0$, the last term is negligible.

Finally, putting $V_k = 0$ and equating eqs. (38) and (40) we obtain

$$V_D = \left(\frac{V_a + \mu V_g}{1+\mu}\right). \tag{45}$$

Thus we say that a diode with a spacing given by eq. (44) and with an anode voltage given by eq. (45) is completely equivalent to the real triode. Thus the anode current in both must be given by Child's law applied to the diode, that is, by

$$I = \frac{2.336\,A \times 10^{-6}\,V_D^{3/2}}{d^2} = \frac{2.336\,A\,(1+\mu)^{1/2}\,(V_a + \mu\,V_g)^{3/2} \times 10^{-6}}{[d_1\,(1+\mu) + d_2]^2} \tag{46}$$

and

$$g = \left(\frac{\partial I}{\partial V_g}\right)_{V_a\,=\,\text{const.}} = \frac{3.504\,A\,\mu(1+\mu)^{1/2}\,(V_a + \mu\,V_g)^{3/2} \times 10^{-6}}{[d_1\,(1+\mu) + d_2]^2}, \tag{47}$$

$$r_a = \left(\frac{\partial V_a}{\partial I}\right)_{V_g\,=\,\text{const.}} = \frac{2.855\,[d_1\,(1+\mu) + d_2]^2 \times 10^5}{A\,(1+\mu)^{1/2}\,(V_a + \mu\,V_g)^{1/2}}. \tag{48}$$

In these expressions A is the cathode area.

We can easily check eqs. (47) and (48) by observing that $gr_a = 1$ as we expect.

These results complete the simple theory for the design of the triode. We notice that, especially for high g valves, the important dimension, which controls g, is the cathode–grid spacing d_1. Thus, to obtain a high g, d_1 should be made as small as possible. However, as we have said before, the work of Bennett and Peterson shows that if d_1 is made less than a, the field on the cathode becomes very non-uniform, all the cathode current originates from regions opposite the centres of the gaps between grid wires and any theory such as the above based on gross equivalent diodes breaks down. Further investigation of these questions is only necessary, however, for valve specialists.

Cylindrical triodes can be handled in much the same way as planar triodes. The results are

$$\mu = \frac{N \ln(R_a/R_g) - \ln\cosh \pi S}{\ln\coth \pi S}, \qquad S = \frac{N_r}{\pi R_g}, \tag{49}$$

$$I = \frac{2.336 \times 10^{-6}\,V_D^{3/2}\,A}{R_k \,.\, R_d \,.\, \beta^2}, \tag{50}$$

$$g = \frac{3.504 \times 10^{-6} V_D^{3/2} A}{R_k . R_d . \beta^2} . \frac{\mu}{1+\mu}, \tag{51}$$

$$r_a = \frac{2.855 \times 10^5 R_k . R_d \beta^2 (1+\mu)^{3/2}}{V_D^{1/2} . A}. \tag{52}$$

Here the notation is as follows: r is the radius of grid wire; R is the mean radius of grid (suffix g), anode (a), cathode (k) and equivalent diode (d); N is the number of grid wires; $V_D = (V_a + \mu V_g)$; and β is Langmuir's β for cylindrical structures.

R_d is defined in this case by

$$\ln (R_d/R_k) = \frac{1}{1+\mu}\left[(1+\mu)\ln (R_g/R_k) + \ln(R_a/R_g) + \left(\frac{\mu+2}{N}\right)\ln \cosh \pi S\right] \tag{53}$$

8. APPROXIMATIONS FOR REAL TRIODES

Frequently, the grids used in small receiving valves are neither plane nor cylindrical and sometimes they are made of mesh rather than of simple wires. Jervis[(12)] gives an expression which accounts for departures from the planar or cylindrical form. It is

$$\mu = \mu_p - K(\mu_p - \mu_c). \tag{54}$$

Here μ_p is planar, μ_c is cylindrical, and K is the number between 0 and 1 depending on the particular shape. His paper should be referred to for details.

Kusunose[(13)] has shown that, for grids of mesh, etc., S can be replaced by the optical opacity of the grid. Thus, for a cylindrical mesh grid in which the meshes are square of side h and the wire thickness is r, N is defined as the active length of grid wire per unit area, i.e.

$$N = \frac{4\pi R_g}{h}(1 - r/h) \tag{55}$$

and

$$\pi S = \frac{Nr}{R_g}. \tag{56}$$

9. MULTIGRID RECEIVING VALVES

In multigrid valves such as screen–grid valves and pentodes, further grids are introduced for various purposes. In the screen–grid a positive grid is introduced between the grid and the anode. This grid is worked at a positive potential usually between 50 and 100 per cent of the anode potential. In this valve the anode current is, within limits, independent of the anode potential so that r_a is very high. The mutual conductance still depends mainly on the grid–cathode geometry so that μ_a is also very high. We can represent I_a by an expression of the form

$$I_a = G\left(V_g + \frac{V_2}{\mu_2} + \frac{V_a}{\mu_a}\right)^n. \tag{57}$$

Here μ_2 is the μ of grid two (screen grid), μ_a is the μ of anode $= gr_a$.

Electrostatic calculation of tetrode μ's is only possible under very

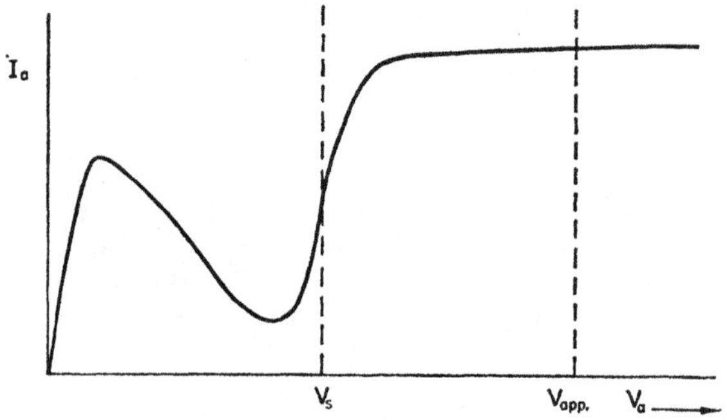

FIG. 7. Idealized tetrode characteristic.

restrictive conditions but approximate results can be obtained by calculating the μ's of two triodes, the first that formed by cathode, grid and screen of the real valve. This is μ_2. Secondly, the μ of the triode formed by grid, screen and anode is calculated. Call this μ_2'. Then

$$\mu_a = \mu_2 \times \mu_2'. \tag{58}$$

Usually, μ_2' will be considerably smaller than μ_2 and often can be calculated by the simplest formulae.

Tetrodes suffer from a severe defect in that if the anode potential swings lower than the screen potential, secondary electrons from the anode can be collected by the screen. The characteristic is of the shape indicated in Fig. 7. In pentodes, a third, open-mesh grid is located close to the anode and is maintained at, or near, cathode potential. This prevents secondary electrons from reaching the screen without interfering very much with the forward flow of current through the device. Again we can approximate the overall μ by breaking up into three triodes, the new one being defined by screen, suppressor grid and anode. The overall μ and r_a are thus even larger than in a tetrode. The suppressor must clearly produce a retarding field at any point on the anode surface, so that the distance between suppressor wire centres should be about equal to the suppressor–anode spacing. It should be transparent so as to reflect few incident primary electrons.

The most important parameter of a receiver valve is the gain × bandwidth product. This is usually defined as the maximum product of gain × bandwidth which can be achieved using the two of the valves under consideration coupled by a low Q resonant circuit. It is then readily shown that (ref. 7, p. 285)

$$\text{gain} \times \text{bandwidth} = g/2\pi C_T, \tag{59}$$

where C_T is the sum of input and output capacitances. More complicated interstage networks can improve on the value given by (59) but this does not invalidate the use of (59) as a criterion of goodness. Thus, the valve designer is interested in obtaining the best value of g/C_T and not merely biggest g. Considerations given in ref. 7 (p. 288 *et seq.*) show that the functional form of this is

$$g/C_T = K\left(\frac{I_c}{d_1}\right)^{1/3}\left[1 - \frac{V_s}{\mu_{12}}\frac{d\mu_{12}}{dV_g}\right], \tag{60}$$

where μ_{12} is the μ of screen grid through control grid and the other symbols are as before. This expression shows that the cathode area does not appear directly, but cathode loading clearly influences the cathode area for a given I_c.

10. ELECTRON-OPTICAL PRINCIPLES IN VALVES

Electron-optical principles are not utilized to any large extent except in the design of transmitter and special valves. In receiver valves the dimensions allowed are so small that the use of lens design is impractical. However, in transmitter valves electron-optical ideas are beginning to be used and R.C.A. have designed high powered screen-grid valves in which electron-optical beam formation was used to reduce the screen current and to minimize reflexion from the suppressor. These valves exploit the ideas on beam formation described in an earlier section on heavy current beams. However, the ideas of field calculation and formation discussed in the earlier part of this article are of the greatest possible value in, at least, the preliminary stages of an electron-optical design and they should be viewed in this light.

REFERENCES

1. Beck, *Thermionic Valves*, Chap. IV, Cambridge U.P., 1953.
2. Lotus and Gunther, *Use of Digital Computers in Electron Beam Technology*, Alloyd Corp., 1961.
3. Southwell, *Relaxation Methods in Theoretical Physics*, Oxford U.P., 1946.
4. Bennett and Peterson, *Bell Syst. Tech. J.* **28**, 303 (1949).
5. Noether, Article in *Differentialgleichungen der Physik*, ed. Reimann Weber, Vol. 2, F. Vieweg, Braunschweig, 1935.
6. Langmuir, *Phys. Rev.* **21**, 419 (1923); Langmuir and Compton, *Rev. Mod. Phys.* **3**, 209 (1931).
7. Beck, *Thermionic Valves*, Chap. 6, Cambridge U.P., 1953.
8. Kleynen, *Philips Res. Rep.* **1**, 81 (1946).
9. Van der Ziel, *Philips Res. Rep.* **1**, 91 (1946).
10. Vodges and Alder, *Phys. Rev.* **24**, 683 (1924).
11. Dow, *Fundamentals of Engineering Electronics*, 2nd ed., John Wiley, 1952.
12. Jervis, *Electronics*, **12**, 1945 (1939).
13. Kusunose, *Proc. I.R.E.* **17**, 1706 (1929).

Printed in Dunstable, United Kingdom

91338436R00121